The Nuclear Economy

To Stephen,

Zach
Maitzen

Cover: A hand holding enough yellowcake uranium to provide the average American consumer with a *lifetime* of electric energy (see page 113).

The Nuclear Economy

Why Only Nuclear Power Can Revitalize The Economy And Environment

ZACHARY MOITOZA

Copyright © 2009 by Zachary Moitoza.

Library of Congress Control Number:		2009907898
ISBN:	Hardcover	978-1-4415-6127-5
	Softcover	978-1-4415-6126-8

All rights reserved. No part of this book may be reproduced or transmitted in any form or by any means, electronic or mechanical, including photocopying, recording, or by any information storage and retrieval system, without permission in writing from the copyright owner.

This book was printed in the United States of America.

To order additional copies of this book, contact:
Xlibris Corporation
1-888-795-4274
www.Xlibris.com
Orders@Xlibris.com

Contents

Introduction 9

Chapter 1	The Global Economic Crisis	13
Chapter 2	Global Peak Oil	19
Chapter 3	Global Warming	29
Chapter 4	The Laws of Thermodynamics	37
Chapter 5	The Quest For Energy	40
Chapter 6	The Steady-State Economy	48
Chapter 7	Electricity	55
Chapter 8	Solar	58
Chapter 9	Wind	65
Chapter 10	Wave and Tidal	69
Chapter 11	Geothermal	70
Chapter 12	Hydroelectric	72
Chapter 13	Biomass	75
Chapter 14	Hydrogen	80
Chapter 15	Energy Efficiency	83
Chapter 16	Nuclear Fusion	86
Chapter 17	Methane Hydrates	88
Chapter 18	Thermal Depolymerization	89
Chapter 19	Space-Based Solar Arrays and Positron Antimatter	91
Chapter 20	Natural Gas	93
Chapter 21	Coal	96
Chapter 22	Phase Five	103
Chapter 23	New and Clear Nuclear	112
Chapter 24	Weapons Proliferation	119
Chapter 25	Waste	125
Chapter 26	Cost	129
Chapter 27	Radiation Controversy and Nuclear Plant Safety	134

Conclusion 143
Web References 147
Glossary 153
Index 169

To Mom and Dad

Introduction

All truth passes through three stages: first it is ridiculed. Second, it is violently opposed. Third, it is accepted as being self-evident.

—Arthur S. Schopenhauer

Dear Reader,

Civilization as we know it is about to change dramatically. This is not reason for alarm; indeed, we should consider ourselves lucky to be living during a time period that historians will one day look upon as a defining event in human history. It will be seen as a major turning point that led to an even greater, more prosperous age for all. For just as historians have named past ages the "Stone Age," "Bronze Age," and "Iron Age," one day they will likely refer to our present time period as the end of the "Fossil Fuel Age." As destined by fate, this pivotal turning point in human history will be replaced by a post scarcity age of advanced, stunningly safe and efficient nuclear "fast reactors," as the age of fossil fuels ends this century.

These are not good times for the United States or anywhere else in the world. There is skyrocketing unemployment, dwindling economic activity, environmental collapse, volatile energy prices and few answers that make sense. Shortly before the economy collapsed in 2008, television commercials promising alternatives like renewable energy and hydrogen became common. But we're still driving fossil fuel powered vehicles, heating our homes on fossil fuels, and indeed running our entire economy on fossil fuel energy, at least what's left of it. Why is this so even after fully eight presidents have been seeking energy independence, from Nixon to Obama? This book was written to provide you, dear reader, with the whole story. How we got here. And how we're getting out.

The first section of the book, "Converging Catastrophes," explains the problems that need solving. Understanding the implications of the massive problems we face is a major part of realizing why nuclear power is needed. Part II, "The Rules of the Game," describes the importance of energy as the basis of prosperity, and why we will always be seeking more energy based upon the laws of physics. Part III, "The Alternatives," explains how none of the much-hyped alternatives to oil will work, except of course for nuclear fission. Understanding this is perhaps the most important concept to grasp, even more important than understanding the many virtues of nuclear itself. And finally, in part IV "The Nuclear Economy," the energy source that will rescue the world from Armageddon gets its own extensive discussion. The truth is that the main problems people associate with nuclear have all been solved using current technology: proliferation, waste, cost, safety, and availability of uranium resources. The few myths about nuclear power with a shred of truth are relevant only to dated technology, and this absolutely incredible energy source is quite literally destined to save the world.

PART I

Converging Catastrophes

The species Homo sapiens is not going to become extinct, but the subspecies Petroleum Man most certainly is.

—Colin Campbell, founder of the
Association for the Study of Peak Oil

Chapter 1

The Global Economic Crisis

Oil is currently the single most important commodity of the global industrial economy, and an understanding of how the global economic crisis unfolded requires an understanding of how vital oil's role is, as the basis of prosperity. The world is today "flat," as Thomas Friedman put it in his book "The World Is Flat," solely because of oil-powered technology providing some truly remarkable services. Industry, agriculture, and especially trade and transport are fueled by oil. As the basis of our transport system, the ships and jet airplanes that crisscross the globe and make the world "flat" are oil fueled. The nearly one billion cars and trucks in use around the world that get people to work on a daily basis and allow trade and commerce form an important foundation of modern civilization. Fossil fuels are the largest industry in the world at over $4 trillion annually; agriculture—itself dependent on fossil fuels—is second.

Fossil fuels provide 85 percent of U.S. energy consumption: 40 percent oil, 23 percent natural gas, and 22 percent coal. Globally, oil is 43 percent of overall energy consumption, but powers 95 percent of transportation. As the basis of transport and other vital economic activities, oil is not just the single largest energy source, but also the most vital, and serves as a major source of economic expansion. Economic growth is tied very closely to increased energy consumption. Following World War II, economic growth in the United States was 7 percent per year, meaning the economy doubled every ten years. Oil consumption rose at the same rate, 7 percent per year, and electricity consumption at an even faster rate. Without energy, we quite simply don't have an economy. And more economic growth will inevitably mean more energy use.

Why oil? What's so special about it? Surely there are alternatives? There are many, but we don't really have any ready-to-scale alternatives that share oil's high energy density, portability and energy return on energy invested (EROEI). At least not right now, while sharing all of those characteristics simultaneously. For example, you can't run an airplane on electricity. Uranium has a vastly higher energy density than oil, but not the same level of portability. One ton of uranium contains as much energy as two million tons of oil, but based upon its physical properties uranium cannot be used to directly power small machines like cars or tractors. Uranium can be used to generate electricity, but battery technology still needs to improve to store that energy for electric cars. Electric cars are still more expensive than internal combustion engine vehicles, and batteries still take longer to charge than a gasoline fill-up.

As such, oil is not only a primary source of energy, it's also energy storage and transport. A single barrel of oil contains 5.8 million BTUs of thermal energy, the energy equivalent of 25,000 hours of human labor. You can fill your car up with gas in three or four minutes and take yourself three or four hundred miles. It is in liquid form and flows to fill the tank. You can use a little gas one day, and the rest waits in the tank until later. Oil is an extremely versatile, efficient carrier of energy. Oil is what gets stuff done the way our economy is currently set up. If oil use is cut off, we're in trouble.

The United States consumes 400 million gallons of gasoline a day. Even when gas was just $2.50 a gallon, Americans spent a billion dollars a day on gasoline. So, a price increase to $5 a gallon would equal an extra billion dollars a day in wealth transfer, and $7.50 a gallon would equal *an extra two billion dollars a day*—or $2,000,000,000. Most of this wealth would be transferred out of the country, since the United States imports two-thirds of its oil. An oil shock can therefore devastate a modern economy since in order for economic activity to continue, huge amounts of wealth must leave importing nations as they purchase oil at high prices. If people respond by cutting back, demand destruction occurs, and economic activity shrinks.

Four out of the past five recessions followed oil shocks (1). The first happened after the Saudi oil embargos of 1973-74. The second after a fall in Iranian oil production after the revolution of 1978. These were both deep and long-lasting recessions. And the third in 1991, after Saddam Hussein invaded Kuwait and set fire to Kuwaiti oil fields. The fourth recession occurred in 2008, shortly after oil prices reached $147 a barrel on the New York Mercantile Exchange on July 11. The economy was still humming along relatively smoothly until a few months after July 11, then it crashed. To put $147 a barrel into perspective, Osama Bin Laden stated in 1998 that his goal to cripple the U.S. economy was to have oil prices reach $144 a barrel (2).

The Nuclear Economy

According to Bin Laden, Americans "have stolen $36 trillion from Muslims" by purchasing oil at low prices. At the time, oil was trading at $11 a barrel on the NYMEX—less than one-thirteenth the July 11, 2008 high.

High oil prices not only heavily tax economic activity, but transfer wealth out of the country to nations that often don't readily spend that wealth. According to Jeff Rubin, chief economist at CIBC world markets,

> Oil shocks create global recessions by transferring billions of dollars of income from economies where consumers spend every cent they have, and then some, to economies that sport the highest savings rates in the world. While those petrodollars may get recycled back to Wall Street by sovereign wealth fund investments, they don't all get recycled back into world demand. The leakage, as income is transferred to countries with savings rates as high as 50 percent, is what makes this income transfer far from demand neutral (1).

Isn't it surprising how the media never mentions the high cost of oil in 2008 as being the cause of the current "Great Recession?" Instead, we are led to believe that symptoms like falling real estate prices and mortgage defaults are the cause. Rubin doesn't feel that plummeting real estate prices are the cause of the recession because the geography doesn't make sense. According to Rubin, "How could real estate prices in Cleveland cause a recession in Japan and the Eurozone?" (1). The number of dollars involved in the oil price shock was vastly greater than any of its resulting symptoms, and indeed, had an effect on the entire world since oil is a global commodity in our presently "flat" global economy. According to Rubin:

> By any benchmark the economic cost of the recent rise in oil prices is nothing short of staggering. A lot more staggering than the impact of plunging housing prices on housing starts and construction jobs, which has been the most obvious break on economic growth from the housing market crash. And those energy costs, unlike the massive asset write downs associated with the housing market crash, were born largely by Main Street, not Wall Street, in both America and throughout the world (1).

Oil, not subprime mortgage defaults or crashing housing markets, was the root cause of the collapse of 2008. As such, alternative energy will be needed to fix the crisis. While there are no ready-to-scale alternatives to oil, long-term shifts away from oil, such as deployment of electric vehicles to replace the internal combustion engine or electric space heating in replace of heating oil, could go a long way in putting downwards pressure on high oil prices. The time to start, obviously, is sooner rather than later.

The way the modern banking system is organized reveals that *it has gotten used to steady economic growth fueled by cheap energy*. Over the past two decades, economic growth has been a steady 2 percent to 3 percent per year, which means economic activity (and, consequently, oil consumption) doubles between every thirty-six to twenty-four years, respectively. Given this reliable rate of steady growth, easy credit was possible. This concept is key to understanding how economic contraction, resulting from high oil prices, resulted in mortgage and loan defaults.

Economic activity was, in effect, used as collateral. As world-respected petroleum geologist Colin Campbell has eloquently put it, "banks lent more than they had on deposit, confident that tomorrow's expansion was collateral for today's debt" (3). When the economy grows, GDP expands, and a larger economy can afford more debt. If the economy starts to contract, people who otherwise would statistically have been able to afford a risky loan suddenly can't.

Banks lent out six to nine dollars for every dollar they actually had on hand in the form of physical cash. By charging high-risk, high-interest prices, what banks are effectively expecting is high economic expansion—the guarantee that you'll be able to make more than you borrowed and pay down your debt plus high interest, thus allowing money to be made by banks. Therefore, by making more than you borrowed through economic expansion, you can pay off your high-risk debt. As soon as the economy began to contract due to the 2008 oil shock, the house of cards crumbled, as banks didn't have the actual money on hand to back their debts, which had fallen through.

Furthermore, without extensive economic growth, U.S. national GDP won't become large enough to support the national debt. If the national debt becomes too large relative to GDP output it ceases to be sustainable. GDP, or Gross Domestic Product, is tied closely to energy consumption used for manufacturing and transporting goods. Energy consumption, GDP, and economic growth are essentially one and the same. If economic activity isn't growing, then debt becomes more difficult to pay off, since its relative size to national output remains high. The trillions in debt the government is amassing can't possibly be paid off without more economic growth or greatly reducing the standard of living of the American people. Energy resources will be vital in allowing GDP to continue to expand to help pay down the national debt.

After the oil shock of July 11, oil prices moderated for about two months. Then, from September 22 to December 22, prices plunged 74 percent to nearly $30 a barrel. As the economy contracted due to the high cost of oil, demand destruction ensued, and oil prices collapsed. During the oil shocks of the 1970s, an inability to meet demand by as little as four percent caused prices to quadruple. Likewise, a drop in demand of a few percent can also take tremendous pressure off the price of oil. The effects on the economy during the period were marked: on September 29, the stock market plummeted 777 points in a single day. By early 2009, the stock market lost more than half its value from all-time highs.

Global financial services firm Lehman Brothers filed for Chapter 11 bankruptcy on September 15, 2008, after suffering massive losses in stock and devaluation of assets by credit rating agencies. At the time, this was the largest bankruptcy in U.S. history. Global Insurance corporation AIG, once the eighteenth largest public company in the world, was taken off the DOW on September 22, 2008. The corporation suffered from a liquidity crisis when its credit ratings were downgraded below AA levels in September 2008. $182 billion in subsidies were not able to save the corporation. Banking and lending practices that worked during times of economic growth suddenly failed, and the resulting entanglement proved too difficult to recover from.

By June 1, 2008, General Motors, the world's second largest automaker, filed for bankruptcy protection. The company was previously the global sales leader for seventy-seven consecutive years from 1931 to 2007. On June 8, the company was removed from the DOW. The years of easy credit, buy-now-pay-later were over. So was the era of the Hummer, which was sold to a Chinese company shortly after the bankruptcy. Perhaps the company could have saved itself if it made smaller, more fuel-efficient cars like competitor Toyota, which penetrated the market with gas-sipping cars after the oil shocks of the 1970s. If you don't learn from your mistakes, you're doomed to repeat them. In the 1950s, people used to say "what's good for GM is good for America." Hopefully, the United States won't go bankrupt as well, but judging by the financial situation in California, that possibility shouldn't be ruled out.

On June 5, 2009, the Bureau of Labor statistics reported U.S. unemployment had reached 9.4 percent according to the U-3 definition often cited by the media. According to the Bureau's U-6 definition, unemployment had reached 15.9 percent. However, had unemployment statistics been computed by the same methodology used into the early Clinton administration, unemployment would be over 20 percent (4). Current U.S. government economic data and reporting, as well as certain private sector numbers, is flawed and gives an unrealistic assessment of underlying economic and financial conditions. About 600,000 jobs a month had been shed since the oil crash, but those "employed" were also working shorter hours, accepting less pay, discouraged to the point that they had given up seeking work, or seeking work for the first time and hadn't yet been laid off. The U-3 statistics cited by the government don't reflect these individuals. In April 2009, 47.1 percent of all people collecting state unemployment exhausted the usual maximum of twenty-six weeks of benefits without finding work according to the Bureau of Labor Statistics (5). The era of cheap oil was over. It was the 1970s all over again, on a global level.

As the economy recovers, oil prices will once again ascend to new highs, and the record oil prices will result in a subsequent crash and then even less economic activity. As the economy continues to contract, the down slope will have highs and lows like a rippling effect, rather than simply being a steady contraction. The economy will shrink a few percent, then start to grow back slightly which will put subsequent pressure on oil prices, crash again due to record prices, grow back slightly while raising oil prices once again to new highs, which will then cause another crash, and so on. The era of price volatility and global contraction had begun. The cause will be slowly declining global oil production. The Earth is finite, and the same barrel of oil can't be taken out of the ground twice.

CHAPTER 2

Global Peak Oil

If four out of the past five recessions were the result of oil shocks, what caused those oil shocks? In 1991, Saddam Hussein set fire to Kuwaiti oil fields. But the other oil shocks were different. They were due to a phenomenon known as "peak oil," the point when geologic petroleum extraction reaches a maximum and goes into inexorable decline. Due to the liquid properties of oil, reserves don't simply run out all at once, but peak and decline slowly. This peaking effect can apply to a single oil well, an entire nation, or the entire world.

Oil is a finite resource. Like coal and gas, oil is a hydrocarbon formed hundreds of millions of years ago when organic materials like diatoms and plant biomass became fossilized, and under the right geologic heat and pressure conditions of the Earth formed the energy-dense fossil fuels that power the global economy today. Most oil is found around a half-mile below the Earth's surface, where just the right level of heat and pressure existed to form oil over hundreds of millions of years. As such, oil is a precious, single-use inheritance. For while the geologic processes that formed oil may still be in place today, they could not create more oil on a time scale relevant to humans.

In 1859, Edwin Drake struck oil in Titusville Pennsylvania using a steam engine powered drill. His oil well produced twenty-five barrels a day. This oil was at first used for illumination in kerosene lamps, since whale oil was becoming expensive as whales were hunted to near extinction. In effect, exhaustion of whale oil for illumination helped spur oil exploration. By 1920, the United States was producing over a million barrels a day, and most of the oil was used to produce gasoline for automobiles. Ironically, gasoline was previously a by-product of kerosene refinement that seemed to have no use. This follows a pattern of growth occurring when one resource begins to decline, leading to a situation that encourages the invention of a new technology like kerosene lamps or gasoline-powered cars that in turn create demand for even larger energy sources. By 1970, after the massive finds of East Texas, ten million barrels of oil a day were being produced. The United States had become the foremost producer of oil in the world.

However, that ten million barrel a day high of 1970 would be the highest rate of oil production the United States would ever produce. 1970 would be the year the United States peaked in oil production. At first, U.S. oil production began slowly, then skyrocketed as "gushers" were discovered in East Texas, named after their ability to propel oil outwards high into the sky by their own pressure. This easy-to-extract oil led to huge increases in production, the "upslope." Vast fortunes were made, which became a part of U.S. culture as seen in shows like the "Beverly Hill Billies," where a poor family became rich by oil sprouting up in their own backyard. However, eventually the easy-to-access oil was mostly extracted, and it reached a point where a majority of oil fields were in decline. A smaller number of oil fields with rising oil production were offset by the larger number of oil fields with declining production, and an overall "peak" in production of oil had been reached.

In the spring of 1971, the "San Francisco Chronicle" carried a single sentence report, not realizing its full implications: "The Texas Railroad Commission announced a 100 percent allowable for the next month." For the first time in history, U.S. oil production quotas, used to regulate oil's price, were completely lifted. Production was allowed to proceed at maximum. However, inexorable declines in production were coming from simply too great a number of oil wells, and U.S. oil production to this day has never reached the level of the 1970 peak. Soon, power over production quotas would shift from the Texas Railroad Commission to OPEC.

This peak phenomenon was accurately predicted by Shell oil Geologist Marion King Hubbert in a speech he gave in San Antonio Texas before Shell Oil Company, on March 8, 1956. The name of this famous speech was "Nuclear Energy and the Fossil Fuels" (6). M. King Hubbert was able to extrapolate a peak in U.S. oil production between 1965 and 1970 based upon the fact that U.S. discoveries of oil peaked in 1936. Hubbert noted that the rate of oil discoveries tended to follow a bell-shaped curve like the rate of oil production from single wells. By analyzing the rate of decline in discoveries, he could predict how much oil would be discovered in the future, and analyzing current production, estimate the time of the peak. Hubbert was right. Today, U.S. oil production is 4.5 million barrels a day—less than half of what it was in 1970. At the same time, we consume nineteen million barrels a day, with the difference made up mostly from imports, and slightly from liquefied natural gas and biofuels. While peak oil doesn't mean "running out" of oil, it can mean a regime of very high oil prices.

Easier to reach, cheap oil is extracted first on land, near the surface, and under pressure. It is also often "light, sweet" crude, meaning it is low sulfur and easier to refine. The remaining oil, the down-slope when production is falling, is often offshore, farther from markets, and in smaller fields and of poorer quality. After 1970 the United States drilled four times more holes for oil than had been drilled everywhere else in the world combined. We opened up the gulf and the super giant Prudhoe Bay in Alaska. Still, we were never able to get up to the ten mbpd level of 1970. All those efforts simply resulted in a gradual down-slope of harder-to-get-at production. Even with advanced seismic technology and drilling techniques, we peaked. M. King Hubbert published his findings in spite of pressure and ridicule, and for over a decade he was a pariah. Few believed his predictions. But, by about 1973 it was clear that the United States had peaked and production could not be increased. A new turning point in U.S. history was reached, marked by oil shocks and far greater emphasis on foreign oil.

It also takes more time, money and energy to refine and transport oil post-peak. For instance, today it costs $10-15 to produce a barrel of oil in the United States. In Saudi Arabia, it costs $2.50 a barrel, and in Iraq, $1. Saddam Hussein delayed Iraq's oil production, and Iraq will likely be one of the last countries in the world to peak at around eight million barrels a day of light, sweet crude by 2018.

The Bush administration was likely very knowledgeable about peak oil. In August 1999, while chairman of Halliburton, before the London Institute of Petroleum, Dick Cheney gave a speech about the future of oil at the end of the century:

> For the world as a whole, oil companies are expected to keep finding and developing enough oil to offset our 71 million plus barrel-a-day of oil depletion, but also to meet new demand. By some estimates there will be an average of 2% annual growth in global oil demand over the years ahead along with *conservatively* a 3% natural decline in production from existing reserves. That means by 2010 we will need on the order of an additional 50 million barrels a day . . . the Middle East, with two-thirds of the world's oil and the lowest cost, is still where the prize ultimately lies.

The world will very likely not be able to meet its 2010 oil demand, even with oil from Iraq. Global peak oil very likely occurred a couple years prior. Steadily growing demand and steadily falling oil production in some of the biggest, most established oil fields means a regime of very high oil prices going forward. U.S. energy policy is going to have to change dramatically in the years ahead, because getting more oil is no longer a possible solution.

Not only was Cheney very much aware of peak oil, so was Bush's energy advisor, Matthew Simmons. Simmons is CEO of the energy company Simmons & Company, and in 2005, published the book "Twilight in the Desert," which argues that Saudi Arabia is likely reaching a peak in oil production. In the 1980s, Saudi Arabia actually increased its reserves estimates from 160 to 260 billion barrels to allow greater production under OPEC production quotas, as did all other Middle Eastern oil producing nations, except Dubai. Obviously, this makes current reserves estimates, the result of so-called "reserves growth," very unlikely. Basically, the books were cooked. As the world's largest producer, if Saudi Arabia peaks, the world peaks.

Based upon his analysis of 200 technical papers published by the Saudis over the past twenty years, which individually detail problems with specific oil wells, Simmons came to the conclusion that the Saudi oil infrastructure was aging and in danger of reaching a maximum rate of output. Simmons noted that most of the oil in Saudi Arabia comes from six aging oil fields, all over forty years old. Extensive seawater injection has been necessary to maintain current flow rates. As Simmons accurately predicted, Saudi Arabia would be unable to increase production enough to meet global demand a few years into the future, and the "world economy will be confronted with a major shock that will stunt economic growth, increase inflation, and potentially destabilize the Middle East." That's exactly what happened. And you thought that you could trust the media to keep you well informed. As Mark Twain put it, "If you don't read the newspaper you are uninformed, if you do read the newspaper you are misinformed."

Another important consideration is the amount of energy required to extract, transport and refine oil. The gushers of East Texas in the 1930s had an energy return on energy invested (EROEI) of 100 to 1. In other words, it took one barrel of oil to get 100 barrels of oil. Today, offshore oil extracted using horizontal drilling techniques has an EROEI of only five to one. So, you get a smaller oil profit. When an oil field is considered depleted there is still oil in the ground, the EROEI has simply fallen to 1:1. If it takes a barrel of oil to extract a barrel of oil, why bother? If you need to use as much energy to power the machinery involved in extracting, transporting, and refining the oil as you get out of it, there is no economic significance in further exploitation.

EROEI can sometimes be important in judging the price and efficiency of an energy source. The only other known energy source capable of an EROEI as high as an oil gusher is nuclear fast breeder reactors, like the Super Phoenix built in France in 1984. Newer versions, like the Integral Fast Reactor, are even more efficient. M. King Hubbert felt that nuclear energy would one day replace the fossil fuel age, as he wrote in his speech "Nuclear Energy and the Fossil Fuels,"

> Ever since the explosion of the first nuclear bomb over Hiroshima in 1945, there has been spectacular evidence that the tremendous store of energy contained within the nucleus of certain unstable atoms can at last be released. The discovery, exploitation, and exhaustion of the fossil fuels will be seen to be but an ephemeral event in the span of recorded history. There is promise, however, provided mankind can solve its international problems and not destroy itself with nuclear weapons, and provided the world population (which is now expanding at such a rate as to double in less than a century) can somehow be brought under control, that we may at last have found an energy supply adequate for at least the next few centuries of the "foreseeable future" (6).

Hubbert calculated that there was enough uranium and thorium to last 5,000 years. For instance, uranium exists in granite at four parts per million, and thorium thirteen parts per million, which if used fully would make granite have fifty times the energy density of coal (6). In fact, as Hubbert noted, a single gram of uranium contains as much energy as three metric tons of coal, or thirteen barrels of oil. The fissioning of a uranium atom unleashes 200 million electron volts—fifty million times the energy of the three or four electron volts released by a carbon atom. Today, technology exists to extract uranium from seawater for only $200 a pound, the energy equivalent of gasoline at a tenth of a cent a gallon. Not only did the speech "Nuclear Energy and the Fossil Fuels" accurately predict the oil peak, it may well have predicted the future energy destiny of mankind.

While Hubbert accurately predicted the peak in U.S. production in his speech "Nuclear Energy and the Fossil Fuels," he later inaccurately predicted a global oil peak for the year 1995. The reason was that global discoveries for oil didn't peak until 1966, so it was impossible at the time to estimate the size of the world's original oil endowment. We now know the answer—about two trillion barrels. Of the two trillion, just over one trillion have been used, and the world has very likely already peaked in 2008. Given that thirty-four years passed from the 1936 U.S. peak in discovery to the 1970 U.S. peak in production, 2000 would be an approximate estimate for a global peak based upon the 1966 global peak in the rate of discoveries. The world now uses six barrels of oil for every one barrel it discovers. However, improved drilling techniques and unconventional oil resources likely delayed the global peak to 2008, which really has been on somewhat of a bumpy plateau ever since 2005, right about on schedule (7).

Of the sixty-five largest oil producing countries in the world, up to fifty-four have passed peak (7). Indonesia in 1997, Australia in 2000, the U.K. in 1999, Norway in 2001, and Mexico in 2004. The U.K. and Norway peaked after their super giant North Sea oil field peaked. Along with Prudhoe Bay in Alaska, the North Sea oil field helped end the oil shocks of the 1970s. The West had finally gained temporary relief from Middle Eastern oil shocks by opening those super giant fields in the 1970s, but their peaks in the 1990s once again shifted influence to the Middle East. Mexico peaked following the peak production of its super giant oil field Cantarell in 2003, previously the second largest oil field in the world after Saudi Arabia's Ghawar. In 2000, nitrogen injection techniques were used to raise Cantarell's production, which reached 2.1 million barrels a day in 2003. By mid-2009, production at Cantarell had plummeted to just 700,000 barrels a day.

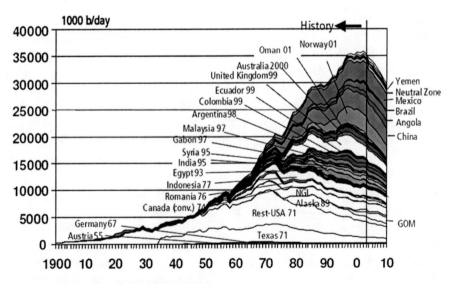

The global peak in conventional oil production was almost certainly in July 2008 at 74.82 million barrels a day (8). By May 2009, it had fallen to just 70.8 million barrels a day (8). Slow declines are expected in world oil production through 2010, as OPEC production increases while non-OPEC production decreases. By 2011, declines may likely accelerate to over 3 percent per year, as production continues to fall in the UK, Norway, and Mexico. Saudi Arabia's massive Ghawar field likely peaked in 2005, and may now be in steep decline, although the Saudi government keeps official production figures a state secret (8). Matthew Simmons was able to uncover some amazing secrets through his detective work. Rising oil production from Iraq, the Arctic, and Brazil's Santos Basin will help keep the decline from being steep over the next decade, but as supply falls and demand continues to rise a regime of very high oil prices and regular oil shocks will become the norm.

In total, three peaks likely occurred: one in 2005 at 74.24 mbpd, a second in December 2005 at 74.16 mbpd, and a third in July 2008 at 74.82 mbpd. As such, the world reached somewhat of a plateau, followed by an eventual peak (8). According to some organizations like the International Energy Agency (IEA) based in Paris, production is measured in "all liquids," which includes natural gas and coal-derived liquids and biofuels. In that case, the peak was also in July 2008 at 87.9 mbpd (8). That number may forever live in infamy.

According to a report issued by the IEA in February 2009, by 2010 oil prices may reach new highs and put global growth at risk (9). Nobuo Tanaka, head of the IEA, said "When demand returns a supply shortage could appear. We are predicting that this shortage could occur in 2013" (9). Of significance was the IEA's official finding that of the 800 largest oil fields in the world, 580 were in decline. Only tiny deposits and unconventional oil were keeping production at a plateau or slight decline.

Every year, Americans on average consume four tons of coal, twenty-five barrels of oil, and seventy-three thousand cubic feet of natural gas. If we were actually able to get all our consumption domestically without having to import, and geologic constraints didn't cause a peaking effect, total conventional U.S. oil reserves would last just three years, and gas reserves just twelve years. Only coal is relatively abundant at 232 years. ANWR, the Arctic National Wildlife Refuge, among the largest of oil fields left untapped in the United States, contains ten billion barrels of recoverable oil—or the amount of oil the United States consumes in a little over a year. In reality, if ANWR were tapped today, production would likely peak at about 875,000 barrels a day in fifteen years, or about four percent of present oil consumption rates, *at peak*. ANWR could make the downslope following the peak slightly easier, but would hardly be a solution. And, as House representative Roscoe Bartlett has said, if we open up ANWR tomorrow, what do we do the day after tomorrow? There will be a day after tomorrow. ANWR should be left untouched so that the grandkids inherit a little oil. The time to start moving away from fossil fuels is now.

While conventional reserves of domestic oil and gas are mostly exhausted, there are still theoretically huge reserves of "unconventional" resources, like the tar sands in Alberta Canada and Shale Oil in the American West. However, these resources are extremely dirty, and most importantly, cannot be produced quickly enough to keep oil production rising. According to U.S. Congressman Roscoe Bartlett, in a Government Accountability Office speech he gave on the floor of the U.S. House of Representatives about the Canadian tar sands,

> They are now aggressively exploiting those fields. They have a shovel that lifts 100 tons at a time. They dump it into a truck that hauls 400 tons, and they haul it to a big cooker... and they are producing about a million barrels a day. But this is just barely over 1 percent of the 84 or 85 million barrels a day that our world produces and our world consumes. And they are using enormous amounts of energy, from what we call stranded natural gas (10).

The tar sands aren't even liquid oil, but solid materials with bitumen absorbed inside. It takes two tons of tar sands to produce one synthetic barrel of oil. This must be laboriously mined, cooked, and treated with freshwater to become oil. The EROEI is very low, around 1.5:1. This is only feasible using cheap natural gas located nearby to provide the energy to cook the tar sands, which won't last much longer (10). A nuclear plant has now been proposed to provide the heat energy to keep the project going. Most importantly, it is unlikely that the project will ever be able to produce oil quickly enough to counter falling production of easy-to-extract, high EROEI oil: the tar sands today produce barely over one percent of world oil production since their refinement is so laborious and cumbersome. There are also major environmental concerns, such as high associated CO_2 emissions and destruction of more than a million acres of boreal forest, now replaced with uninhabitable toxic wasteland due to runoff from the refinement process.

While oil extraction from the tar sands is itself very destructive and cumbersome, shale oil extraction from the American West is even worse. Nuclear power would also be needed to produce liquid oil from "shale oil" in Colorado, which has to date produced absolutely zero oil. Exxon tried processing shale oil in 1982, and gave up after one day. Shale oil is simply solid rock with kerogen, a mixture of organic chemical compounds found in sedimentary rocks, absorbed inside. Turning this into oil requires insitu temperatures of 8500°F followed by treatment with hydrogen to make it flow, then flash-freezing back to low temperatures (11). The process would likely require several nuclear plants and diverting the entire Colorado River to provide enough hydrogen (11). According to Shell geologists, oil prices would have to reach $300 a barrel before shale oil extraction could become economical.

The EROEI for shale oil would also likely be negative, meaning that it would take more energy to refine it than is contained in the end product (11). As such, only nuclear power could economically provide the heat energy needed for the process. Since we still don't drive electric vehicles and may be faced with a liquid fuel crisis, in an act of desperation we may effectively try to turn uranium into oil at an energy loss. Again, even if we did embark upon such an endeavor in absolute desperation, the amount of oil produced, at absolute maximum a few million barrels a day, wouldn't fully mitigate declines of easy-to-extract oil. And the Colorado environment would suffer tremendously. Hopefully, we won't be foolish enough to use this stuff for anything more than asphalt, which has a much less energy and water-intensive refinement process.

The U.S. Department of Energy commissioned study "Peaking of World Oil Production: Impacts, Mitigation and Risk Management" was released in early 2005, known as the Hirsch report after primary author Robert L. Hirsch (7). The report warns,

> As peaking is approached, liquid fuel prices and price volatility will increase dramatically, and, without timely mitigation, the economic, social, and political costs will be *unprecedented*. Viable mitigation options exist on both the supply and demand sides, but to have substantial impact, they must be initiated more than a decade in advance of peaking (7).

If M. King Hubbert was right about the U.S. oil peak in 1970, and predicted a global peak around the turn of the millennium, why didn't any of our leaders sound the alarm and do something about this issue? Since the United States is a microcosm of the world, wouldn't you hope somebody would have made the global peak a priority? Where was the leadership on peak oil over the past three decades?

Hirsch later remarked, "The world should urgently begin spending $1 trillion a year in crash programs preferably *two decades before* the peak" (7).

The world has almost certainly already reached peak oil. "Liquid fuel prices and price volatility" have already arrived. Essentially nothing has been done, even though preparation should have begun decades earlier. No new nuclear plants have been built in the United States start to finish since 1973. Most of the 104 vintage nuclear plants now operating in the United States were built during the two decades following the construction of the 1957 Shippingport reactor, and now the United States will have to build about 500 more of the biggest two-gigawatt nuclear plants at an even faster rate, to provide the terawatt of energy it will need mid-century—not to mention electric cars, electric trains, and homes retrofitted with electric heaters, etc. The question now is only how difficult the transition will be. How *unprecedented* will the "economic, social, and political costs" become? Will the United States still be able to afford these costs given future economic collapses? Great hardship likely lies ahead, which will become much worse than anything experienced during the Great Depression. Only one thing is certain: rapid expansion of alternative energy will become the single most important priority of the second decade of the twenty-first century.

Chapter 3

Global Warming

Whereas few have heard about peak oil, nearly all have heard about global warming. Few, however, understand the terrifying implications of anthropogenic climate change and the sheer magnitude of greenhouse gases that humans are evaporating out of the ground as fossil fuels deplete. Few are also aware, in contrast, of how tiny our nuclear waste footprint is in comparison with our carbon footprint. As James Lovelock, founder of the Gaia hypothesis that the Earth is a self-regulating ecosystem, puts it,

> An outstanding advantage of nuclear over fossil fuel energy is how easy it is to deal with the waste it produces. Fossil fuel burning produces twenty-seven thousand million tons of carbon dioxide yearly. This is enough if solidified to make a mountain nearly two kilometers high and with a base ten kilometers in circumference. The same quantity of energy if it came from nuclear reactions would make fourteen thousand tons of high-level waste. A quantity that occupies a sixteen meter sided cube (12).

Truly, our nuclear footprint is unbelievably tiny compared to how shockingly massive our carbon footprint has become—a difference by a factor of two million for the same quantity of energy produced. I cannot emphasize enough the implications of that number. If something is two million times as energy dense, it produces two million times less waste. A thousand times a thousand is a million—so a thousand nuclear power plants running for a thousand years would produce as much waste as one coal plant running for less than one year. If someone were to say that nuclear fuel is ten times as energy dense that might sound pretty good, but two million times the energy density truly has profound implications as far as cleaning up the environment. For fossil fuel waste "the solution to pollution is dilution" through massive emissions into the environment. For nuclear waste, the amount produced is so tiny it can be sequestered away with great care. 80 percent nuclear France has the cleanest air of any nation in the industrialized world, and stores all its nuclear waste from decades of operation in one room in La Hague.

Carbon dioxide, a proven greenhouse gas, influences climate by interfering with the natural heat loss of the Earth. As solar energy warms the Earth, some reflects back out, or radiates outwards due to natural cooling. The sheer quantity of CO_2 the world now emits from combustion of coal, oil and natural gas is truly staggering in volume. At the start of the Industrial Age around 1750, the concentration of CO_2 in the atmosphere was 280 parts per million (ppm). Today, it is 388 ppm (13). Nuclear power, in contrast, produces no CO_2 during energy production. According to President Obama's climate stabilization strategy, CO_2 emissions must be reduced 80 percent from 1990 levels by 2050. This would be a five-fold reduction in waste. But because nuclear energy produces about two million times less waste per unit of energy produced, a lifetime of electricity could be supplied to a family of four by a piece of uranium the size of a golf ball.

The fourth assessment report of the Intergovernmental Panel on Climate Change (IPCC), issued in February 2007, said it can now be stated with "very high confidence" that human activity is having an effect on climate. The world's climate has increased about 1°F over the past few decades, and about 2°F over land areas. Further warming is certain due to gases already in the air because of climate system inertia and inevitable additional fossil emissions. The global temperature is at the highest level of the Holocene, the past 10,000 years when civilization developed.

In November 2008, the leading climate scientist at NASA, Dr. James Hansen, wrote an essay to President Obama titled "Tell Barack Obama the Truth: The Whole Truth," explaining the urgency needed for climate stabilization (13). Hansen is the world's preeminent climate scientist. According to Hansen, the effects of climate change are already becoming evident. Mountain glaciers are receding worldwide and will be gone in fifty years at current rates, threatening the freshwater supply of millions of people. Coral reefs, home to a quarter of all biological species, are being destroyed due to increased ocean acidity from CO_2 absorption. Dry subtropics are expanding poleward, increasing drought and wildfires in the Southern United States, Mediterranean, and Australia. Arctic sea ice will soon disappear entirely in the summer, affecting polar bears and Eskimos.

And finally, extreme weather is becoming more frequent, affecting some areas with drought and others with more heavy rain and storms (13). There are some who believe that we must learn to live with these worsening changes since it is too hard to move away from fossil fuels; hopefully they are simply uninformed about atomic energy.

Most alarming about Hansen's report is that climate change will not follow a linear progression, but rather an exponential progression reinforced by positive feedback loops, meaning that if action is not taken soon the climate crisis could become uncontrollable. Like a wildfire just beginning to spread, if CO_2 concentration reaches 500 ppm it may be too late to prevent going over a "tipping point" of uncontrollable warming, leading to unstable ice sheets, sea levels rising out of humanity's control, extermination of a large fraction of species on Earth, and severe exacerbation of climate.

According to Hansen,

> If the process proceeds too far, amplifying feedbacks push the system dynamics to proceed without further human forcing. For example, species are interdependent—if a sufficient number are eliminated, ecosystems collapse. In the physical climate system, amplifying feedbacks include increased absorption of sunlight as sea ice and land ice areas are reduced and release methane, a powerful greenhouse gas, as permafrost melts (13).

Ice is white and reflects sunlight. Melt the ice and less solar energy reflects back out. *There are also huge reserves of frozen methane in the arctic and below the ocean floor, which could be released by further melting—and methane is twenty three times as potent a greenhouse gas as carbon dioxide.* Because CO_2 remains in the atmosphere for 1000 years, quickly reducing CO_2 levels to 350 ppm may be the only way to avert a climate collapse.

It may be too late to do so, even with peak oil slowing down oil consumption. However, according to Hansen, if global coal plants are linearly phased out from 2010 to 2030, atmospheric CO_2 may peak over the next few decades at around 400-425 ppm. Coal is the largest source of CO_2 emissions worldwide. In the United States, over 600 coal plants providing 49 percent of total electricity generation put out 40 percent of total CO_2 emissions—all the cars on the road are 32 percent. While it is too late to prevent further warming, the rapid elimination of coal may be the only way to avoid the 500 ppm tipping point when positive feedback loops kick in and it becomes too late to avoid catastrophe.

This will be no easy task. Coal plants produce reliable, 24/7 power, called "base-load," at low cost, and can be built almost anywhere. China alone is building one to two coal plants a week. Global coal consumption is over six billion tons a year. Hansen recommends the rapid deployment of fourth generation nuclear power plants, which could be ready as early as 2015 (14). The two proposed models are the Integral Fast Reactor and Liquid Fluoride Thorium Reactor. Both models are 100 to 300 times as fuel efficient as today's generation II light water reactors, and can operate for several centuries on uranium and thorium that has already been mined, including existing uranium and plutonium in nuclear waste (13). Both reactors use a proliferation-resistant pyrometallurgical fuel cycle that incinerates long-lived isotopes in spent fuel for more energy, reducing waste isolation time to less than 500 years. According to Hansen,

> It is specious to argue that R & D on 4th generation nuclear power does not deserve support because energy efficiency and renewable energies may be able to satisfy all United States' electrical energy needs. Who stands ready to ensure that the energy needs of China and India will be met entirely by efficiency and renewables? (13)

It is expected that total world energy demand will at least double by 2050 from today's fifteen terawatts to over thirty terawatts. If this energy is supplied by coal, the environmental consequences could be catastrophic. Furthermore, we have yet to demonstrate large-scale, long-term storage of carbon dioxide. As oil production peaks and more electric cars become available, demand for electricity worldwide could grow even faster. The world desperately needs a massive source of reliable, long-lasting, low-pollution energy, and nuclear power is clearly a possible solution. It remains to be seen if so-called "renewable" energies will be able to meet this challenge. And, according to Hansen, dramatic reductions must begin immediately *"if humanity wishes to preserve a planet similar to that on which civilization developed and to which life on Earth is adapted."*

One of the world's foremost proponents of clean energy was Richard Smalley, winner of the 1996 Nobel Prize in chemistry for the discovery of a new type of carbon atom nanotechnology, the "bucky ball." Smalley would routinely give speeches, still available online, that the single most important issue facing humanity this century was finding a new source of abundant, clean, cheap energy. Smalley would assemble a list of problems facing humanity: (1) Energy, (2) Water, (3) Food, (4) Environment, (5) Poverty, (6) Terrorism and War, (7) Disease, (8) Education, (9) Democracy, and (10) Population (15). According to Smalley, all of these problems could be solved if the energy problem is solved. Ocean water can be desalinated and pumped to where it is needed—this is just energy intensive. Food production will require water, now increasingly in short supply around the world as drought worsens and glaciers recede. Even population begins to decline in developed nations, as higher levels of education are obtained and people have access to modern practices like birth control. According to Smalley,

> Energy is the single most important factor that impacts the prosperity of any society. It is impossible to imagine bringing the lower half of the economic ladder of human civilization—about three billion people—up to a modern lifestyle without abundant, low-cost, clean energy (15).

According to Smalley, we must find the "new oil" this century, and fast. It must be a vast new energy source that will replace the fossil fuel energy regimes of the past and usher in a new age of prosperity for all. If that "new oil" is not found, the twenty-first century isn't going to be a very prosperous century. In fact, the twenty-first century may be remembered by historians as the century that civilization as we know it comes to an end.

PART II

The Rules of the Game

The greatest shortcoming of the human species is our inability to understand the compound growth formula.

—Albert Bartlett, professor of Physics,
University of Colorado at Boulder

Chapter 4

The Laws of Thermodynamics

Since Medieval times, inventors have tried in vain to develop a "perpetual motion machine," a source of unlimited industrial energy. Such devices are also referred to as "free energy" devices. The first such recorded attempt was in 1150, when Indian mathematician Bhaskara II described a wheel he claimed would run forever. For centuries, many brilliant inventors devoted their entire lives to such machines, only to fail. Even Leonardo da Vinci made drawings of various machines he hoped would be able to run forever. Such prototypes often involved magnets, and would have a crank to start the machine. However, it would simply stop after the crank was no longer turned. If one pushes a crank to force magnets together, one is just storing the amount of muscle energy he or she expended by moving the crank, so no new energy is produced. Many tried to develop a free energy device, but none succeeded.

It was finally determined by modern science that developing a perpetual motion machine is impossible, which led to the eventual development of the laws of thermodynamics, perhaps the most important principles of modern physics. The first law of thermodynamics states that energy can be changed from one form to another, but it cannot be created or destroyed. The total amount of energy in the universe remains constant, merely changing from one form to another. This is known as the law of conservation of energy. Energy is always conserved in any transaction. If a lump of coal is burnt, no more energy is produced than was stored within the coal itself. The energy simply changed form, from the potential energy of the bonds of the coal's hydrocarbon chains to heat energy. The amount of energy in the universe remained fixed.

The second law of thermodynamics states that in all energy exchanges, if no energy enters or leaves the system, the potential (usable) energy of the state will always be less than that of the initial state. Energy only converts in one direction, from useable to unusable. Whereas "conservation" sums up the first law, "entropy" sums up the second. There is always a tendency for energy to move from a more concentrated to a less concentrated state. Just as how a house of cards will tend to eventually collapse, but will never spontaneously build itself up (without further energy input by a skilled artist), so too will energy follow the laws of entropy, and move from a concentrated to a less concentrated state.

The phenomenon was first observed by German physicist Rudolf Clausius in 1868, and led him to coin the term "entropy." Clausius noticed that in order for energy to be converted into work, there must be a difference in concentration (i.e., difference in temperature) in different parts of a system. Work therefore occurs when energy moves from a higher concentration to a lower concentration, or a higher temperature to a lower temperature. For example, a steam engine extracts useful work from the fact that one part of the engine is very hot while the other is very cold. A hot rock in the summer sun will eventually cool at night as heat dissipates into the air, and energy ultimately radiates back out into outer space. To maintain a constant temperature requires more energy, such as how a person burns food calories to maintain homeostasis. In order to maintain a difference in temperature, more energy must ultimately be applied.

The significance of the laws of thermodynamics is that *we will always be seeking energy*. Since energy cannot be created, and cannot be reused, we will always need an ultimate source of energy to meet our energy needs. This is not true for metals, which leave behind a scrap that can be recycled. Even molecules that break off and disperse may one day make their way into seawater where they can again be extracted—using more energy. However, fuel leaves no scrap. The steam engine requires continuous supplies of coal, for example. The animal continuous food. The plant continuous sunlight. All of the energy in the universe is somehow stored, and can never again be reused after it is released and does work. Heat that does work dissipates until there is no longer a heat gradient, and ultimately radiates back out into the endless depths of outer space.

Another consequence of the laws of thermodynamics is that *renewable energy doesn't exist*. Such a phenomenon would of course violate the second law, for no energy, once used, can be regenerated or be "renewed." There are ultimately only three sources of energy on Earth: the energy stored in fossil fuels within the Earth, the energy stored in nuclear fuels within the Earth, and the influx of solar radiation to the Earth. All other sources are just derivatives of these sources. For example, the differential heating of the Earth stirs up wind. The sun's evaporation of moisture creates rain, which flows into rivers, creating hydropower. Plants capture solar energy through photosynthesis, creating biofuels. Even geothermal power is the result of the radiation given off by abundant uranium and thorium atoms within the Earth, which heat the Earth's core to temperatures hotter than the surface of the sun.

As such, there are really only two ultimate energy sources available to humanity: nuclear and solar. For fossil fuels, given their finite quantities and high levels of pollution, are not sustainable. The main problem with solar energy, however, is that it is extremely diffuse and intermittent. The relentless quest by scientists to harvest vast quantities of this "renewable" energy through solar panels and wind turbines may one day be looked upon in the same way as past inventors' quests for perpetual motion.

Chapter 5

The Quest For Energy

    The sun produces its energy through a process known as nuclear fusion according to the equation E=MC squared, or energy equals mass times the speed of light squared. In other words, there is a huge store of energy in matter. As such, nuclear fuels are about two million times as energy dense as the chemical energy of fossil fuels. The massive heat and pressure of the sun's core fuses a thousand tons of hydrogen into helium a second, creating enough energy in that one second to power modern civilization for a million years. The energy output is so vast the sun is blindingly bright ninety-three million miles away.

    It is this influx of energy that has allowed life on Earth to exist for billions of years. Plants, or producers, capture solar energy though photosynthesis and convert it into a durable hydrocarbon by using the energy to chemically bond carbon and hydrogen. Animals eat plants, or other animals, and release this stored energy through oxidative respiration. As such, most of the biomass on the planet is plant life, a much smaller amount forms herbivores, and an even tinier amount forms carnivores. This is due to the fact that energy is lost as it changes form—an organism uses most of the energy it obtains from the environment to fight entropy and maintain internal homeostasis. If the organism is itself eaten, a smaller amount of energy is available to the next level of the food chain. Eagles, at the top of the food chain, represent far less biomass than mice, which represent a tiny amount of biomass compared to wood. The amount and type of biomass an ecosystem can support is directly proportional to the amount of energy available. In that sense, the ecosystem is a way of capturing and expending energy as it flows through the food chain.

The Nuclear Economy

Life is in essence a quest for energy. As Erich Jantsh remarked in "The Self-Organizing Universe," we seem to live in a universe where orderly structures form whenever there is a flow of energy. Living organisms evolved, in effect, based upon how successful they are at capturing energy. If a fox isn't very good at catching prey, and has to expend more energy from its muscles chasing rabbits than it ultimately obtains by eating rabbits, pretty soon no more fox. There is only so much space available for plants to grow and capture solar energy, only so much plant life available for herbivores to eat, etc. As such, those organisms more efficient and successful at capturing energy were the ones who survived to reproduce, and all life forms on Earth are the descendants of organisms more successful at capturing solar energy. The more energy an organism captures, in turn, the more sophisticated and evolved it becomes. For instance, mammals, one of the higher orders, require much more food energy than bacteria. Sophisticated plants have evolved pollens that spread in the wind, and roots that absorb rain that falls after being evaporated by the sun.

According to cultural anthropologist Leslie White, human culture is no different. According to White's Law, the measure by which to judge the relative degree of evolution a culture has obtained is by the amount of energy it captures. Under White's Law "culture evolves as the amount of energy harnessed per capita per year is increased, or as the efficiency of the instrumental means of putting the energy to work is increased." The technological level of society, as such, is entirely dependant upon the amount of energy it uses. White observed that technology is an attempt to solve the problems of survival, which ultimately means capturing enough energy and diverting it for human needs. Societies that capture more energy and use it more efficiently have an advantage over other societies, and are therefore more advanced in an evolutionary sense.

White described five levels of human society:

(1) Energy is derived from human muscle power.
(2) Energy is derived from domesticated animals.
(3) Energy is derived from domesticated plants.
(4) Energy is derived from fossil fuels.
(5) Energy is derived from nuclear fuels.

According to White, all the human creativity in the world will inevitably come up short in advancing wellbeing absent sufficient energy resources to be captured and harnessed. The claim is sometimes made that the United States needs to embark on an Apollo Program type project to get itself off fossil fuels, but fossil fuels are what powered spaceships to the moon, and forged the steel used to manufacture them. The technological advances we take for granted have essentially just been clever new ways of figuring out how to use fossil fuels, so getting "off" them would require an even better energy resource base. As fossil fuels today deplete and pollute, nuclear fuels, the final stage, will be needed to power civilization. The discovery of nuclear fission truly was the greatest discovery of the twentieth century, the discovery that would allow other advances to become sustainable.

Anatomically modern humans emerged around 50,000 years ago after what anthropologists refer to as a "great leap forward," perhaps due to the perfection of the voice box. As hunter-gatherers, these humans were at the first stage of White's level of civilization. Human muscle power is rated at 1/20th horsepower, or about thirty-five watts continuously. White saw these humans as "power plants" that hunted and gathered wild foods for energy from their environment. Technology of the time period only reflected what could be used to capture and use energy from the environment more efficiently under that type of energy level. Bone needles were used to sew clothing to maintain body heat in cooler climates. Harpoons, spears, and bows-and-arrows leverage energy from human muscle power more efficiently, allowing the arm or fingers to fell large beasts. Stone tools for grinding seeds were used to obtain food energy. As busy nomads, little surplus energy was available to invest effort in elaborate living arrangements or artwork. Only that which could be carried was valued, so civilization did not have the opportunity to progress rapidly.

The next shift up White's ladder occurred around 10,000 years ago in the Fertile Crescent of the Middle East, when humans first domesticated wild plants and animals. This occurred at the start of the Holocene, when climatic changes made the Middle East more fertile. As demonstrated in Jared Diamond's "Guns, Germs, and Steel," the Fertile Crescent was endowed with wild cereals and animals relatively easy to domesticate, which led to a vast rise in human culture. Early hunter-gatherers were able to obtain fifty calories of food energy for every one calorie of muscle energy used to harvest wild wheat—a fantastic EROEI. Wild wheat was itself very easy to domesticate since it evolved to have its seeds shatter and spread in the wind, and a single-gene mutation prevented this from happening, leaving the seeds stranded in their pod waiting to be harvested. Wheat seed is also high in protein, and its shell allows it to be stored over winter months. Barley, goats, and cattle were domesticated in the Middle East as well, leading to a compelling food package. Parts of the wheat plant that weren't edible could be fed to livestock, which in turn would produce manure, which in turn could fertilize more crops. More livestock also meant more food, and more labor for tilling soil or horses for trade and transport.

The Nuclear Economy

Human populations grew in number exponentially since far more food energy was available. Ample harvests led to stored food surpluses, and dense, stratified, sedentary societies. Some people didn't need to farm due to stored food surpluses, and devoted their time to the arts and sciences. The slaves who built the Egyptian pyramids were fed by domesticated wheat and barley that spread from the nearby Fertile Crescent, as were Greek scribes and scholars. Land once occupied by inedible forest was taken over and replaced with efficient domesticated crops, thereby devoting a vastly greater percentage of the Earth's solar influx to human endeavors, and a pastoral culture emerged.

By the 1700s, England was becoming too overpopulated. The population of the world, just three million in 35,000 BC, was now 500 million. Domesticated crops from the Middle East had now spread around the globe. Heating fuel became much more expensive as forests were depleted, and the English population turned increasingly to its abundant coal reserves for home heating. England originally had more coal than Saudi Arabia had oil. However, coal mines tended to flood, and it was difficult to pump water out of the mines. In 1712 Thomas Newcomen invented a coal-powered steam engine for pumping water out of coal mines. The problem, however, was that his steam engine was only two percent efficient, and barely even worked.

In 1769, James Watt greatly perfected the steam engine by improving its thermal efficiency five-fold, after developing a model with extended heat differentials. Soon, coal-fired steam engines were powering industrial factories and locomotive trains. Fossil fuel use in England increased exponentially afterwards, as it later did around the world. A century later, a use would be found for gasoline, previously a useless by-product of kerosene refinement for lamps, with the invention of the internal combustion engine. What we think of as modernity was ushered in. This transition is commonly referred to as the "Industrial Revolution," but the "Fossil Fuels Revolution" would have been a more fitting term. When access to a new energy resource enters a civilization, culture inevitably evolves.

A single barrel of oil contains 5.8 million BTUs of thermal energy, the energy equivalent of twelve people working for a year. The average American today consumes the fossil energy calorific work equivalent of 150 faithful servants toiling around the clock, or 500 people working normal shifts. The energy equivalent of 2,000 men push our cars down the road, and 244 men are at the beck and call of the average industrial factory worker. 700,000 men lift a jet airplane across the sky. Sewing machines manufacture cheap clothing that in the past was made by using needle and thread, and homes once made by hand utilize building materials, industrially manufactured. The average American enjoys a higher standard of living than the wealthiest pharaohs of ancient Egypt, who had to rely on human or animal labor, rather than fossil energy labor. More free time is thus available, for leisure, learning, and the arts. Americans live in what might truly be considered a golden age, due to an incredible and unprecedented access to energy resources.

At 1:50 p.m. on December 20, 1951, the EBR-I Experimental Breeder Reactor became the world's first electricity generating nuclear plant, illuminating four 200-watt light bulbs. General Dwight D. Eisenhower remarked that mankind finally has an energy source other than fossil fuels, and envisioned an age of man powered entirely by nuclear energy. By 1953, the EBR-I was breeding as much plutonium as it was consuming—the closest thing to a perpetual motion machine ever developed. Breeder reactors make use of all of the energy in uranium by converting it into a fissionable isotope of plutonium, Pu-239. Today's more common light water reactors use less than one percent of the potential energy of uranium.

In 1983, Nuclear Physicist Bernard Cohen argued that by extracting uranium from seawater for use in breeder reactors, using a fabric adsorbent submerged system, nuclear power could meet the definition of the commonly used word "renewable" (16). The world, at the time, was only believed to hold enough high-grade uranium ore to fuel light water reactors for a few decades. Today, more uranium has been discovered in Australia and Canada, and there is enough to fuel light water reactors for a few centuries with moderate demand growth. Uranium is as abundant as tungsten or tin, and thirty-five times as abundant as silver. According to Cohen, since breeders are 100 times as efficient, even at $400 a pound uranium extracted from seawater would be the energy equivalent of 1.1 cents per million BTUs if used in breeders. Based upon various geologic uplift and erosion, rivers wash enough uranium into seawater for us to extract 16,000 tons a year indefinitely, or at least until the sun consumes the Earth several billion years from now. This is enough uranium, if used in breeder reactors, to provide about twice the world's total 1983 level of energy consumption for as long as life on Earth is sustainable (16). This could be stretched even further through improved energy efficiency in electric cars or other applications.

The inventor of the light water reactor, Alvin Weinberg, later remarked in his memoirs,

> The breeding ratio of France's sodium-cooled Phenix has been shown to be 1.13, and Admiral Rickover's thorium-based U-233 seed-blanket light water breeder has been shown to have a breeding ratio of around 1.01. These demonstrations of actual breeding have passed rather unnoticed. I regard them as extremely

important, since we can now say with absolute certainty that nuclear fission, based on breeders that burn very low-grade ores, represents an all but inexhaustible source of energy (17).

By breeding more fissile material (fissionable material) than they use, breeders can completely consume uranium and thorium resources, thus making nuclear fission a truly sustainable energy source. While the technology has existed for decades, new designs are extremely safe and proliferation resistant (18). Truly, Leslie White was wise in predicting that nuclear fission power would become the ultimate enabler of human material progress.

A good example is the history of the navy. The sailing ship, which replaced the slave-driven galley of antiquity, was vastly improved by medieval ship builders and became the first machine enabling man to control large amounts of inanimate energy. As such, man transitioned to wind energy away from human muscle power, allowing the exploration of the globe. In the 1800s, the English Navy built steam-engine powered metallic ships fired by coal, which greatly improved the speed and endurance of ships, and allowed them to travel nonstop, regardless of when the wind blows. In 1911, Winston Churchill was given the task of making sure the British Navy was prepared for war, and decided to replace England's coal-powered vessels with oil-powered ships. The superior energy density of oil allowed his naval ships to travel faster and refuel less often, giving the English navy a strategic advantage over WWI German ships. During WWII, diesel powered submarines were implemented that could stay submerged for a few hours while running on battery power, for there is no air underwater for fossil fuel combustion. This transition to ever more energy dense fuels enabled ever more powerful naval vessels, for coal contains as much energy as three tons of wood, and oil as much energy as four tons of wood.

In contrast, one ton of uranium contains as much energy as ten million tons of wood. The world's first nuclear powered submarine, the USS *Nautilus*, was commissioned on September 30, 1954, and ushered in a new era of submerged speeds and endurance. The Nautilus rendered progress made in antisubmarine warfare during WWII virtually obsolete. Radar and anti-submarine technology crucial in defeating submarines during WWII was useless against a vessel capable of moving out of an area in record time at high speed, changing depth quickly and staying submerged for very long periods. On February 4, 1957, the Nautilus logged 60,000 nautical miles of travel underneath the sea, matching the endurance of the fictional Nautilus described in Jules Verne's novel "Twenty Thousand Leagues Under The Sea." Previously, such voyages were only works of science fiction since fossil fuels remove far too much oxygen from the air to be burnt in enclosed underwater vessels.

On August 3, 1958, the Nautilus became the first vessel to reach the North Pole—by traveling under the sea ice. The ship was capable of staying submerged for years at a time. It would take in seawater and boil it to provide the crew with drinking water, and use electricity to split apart water molecules into hydrogen and oxygen to provide the crew with oxygen to breath. The fuel was uranium enriched to 20 percent U-235 in a compact light water reactor, which would release no emissions into the environment, suck no air from the environment, and was completely sealed inside the vessel. The amount of uranium 235 used over the ship's entire thirty-year lifetime of service would fit in a backpack.

A single diesel powered container ship, in contrast, can emit cancer and asthma-causing pollutants equivalent to that of fifty million cars. There are today about 150 nuclear powered ships in operation around the world, mostly submarines, which range from ice breakers to aircraft carriers. A Nimitz-class super carrier has more than twice as much horsepower (240,000 hp or 208 MW) as the largest container ship diesel engines ever built and is capable of continuously operating for twenty years without refueling. The U.S. Navy has today accumulated over 5,400 reactor years of operation with zero reactor accidents and zero pollution emissions. Worldwide, 60,000 premature deaths a year are the result of particulate matter emissions from ocean-going ships, as well as large amounts of greenhouse gas emissions. Switching to nuclear-powered ships for global commerce and travel could go a long way in helping the environment and improving upon ship safety.

Largely accredited with helping the navy achieve this excellent record is the strict safety design standards imposed by Admiral Hyman G. Rickover, "father of the nuclear navy." In 1957, before a banquet of the Annual Scientific Assembly of the Minnesota state Medical Association, Admiral Rickover gave an excellent speech that covered all the themes of peak oil and energy sustainability (19). The speech may one day be regarded as the most prescient speech ever spoken. Some of Rickover's remarks included truly fascinating information about populations and energy:

> Surplus energy provides the material foundation for civilized living—a comfortable and tasteful home instead of a bare shelter; attractive clothing instead of mere covering to keep warm; appetizing food instead of anything that suffices to appease hunger. It provides the freedom from toil without which there can be no art, music, literature, or learning. A reduction of per capita energy consumption has always in the past led to a decline

in civilization and a reversion to a more primitive way of life. For example, exhaustion of wood fuel is believed to have been the primary reason for the fall of the Mayan Civilization on this continent and the decline of once flourishing civilizations in Asia. India and China once had large forests, as did much of the Middle East. Deforestation not only lessened the energy base but had a further disastrous effect: lacking plant cover, soil washed away, and with soil erosion the nutritional base was reduced as well (19).

Rickover obviously must have been aware of the teachings of Leslie White and M. King Hubbert, which were contemporary for the time period. Energy flows through a population determine how advanced that population may become, and the head start Asia had over Europe a millennium ago was ultimately lost due to resource depletion and a resulting decline in civilization into poverty, which still exists in the region to this day. In contrast, England turned to its abundant coal resources when its wood resources began to deplete, allowing per capita energy use to continue to rise. Thus, Europe's fate was very different from those of less fortunate societies like the Maya. However, shifting to fossil fuels has given rise to new problems. As Rickover explains,

> The Earth is finite. Fossil fuels are not renewable. In this respect our energy base differs from that of earlier civilizations. They could have maintained their resource base through careful cultivation. We cannot . . . More promising is the outlook for nuclear fuels. These are not, properly speaking, renewable sources of energy, at least not in the present state of technology, but their capacity to "breed" and the very high output from small quantities of fissionable material, as well as the fact that such materials are relatively abundant, do seem to put nuclear fuels into a separate category from exhaustible fossil fuels.

Economist Julian Simon once remarked that "energy is the master resource." As such, uranium and thorium must be the master resources—the final enablers of material progress. Rickover was well aware of the virtually limitless energy nuclear fuels could provide, and the limitations of finite fossil fuels. Renewable energy, in his view, could fashion only a very small part of our energy needs—at the most 7-15 percent. To this day, only 6 percent of U.S. energy consumption is "renewable," mostly hydro (20). Wind and solar combined are just 0.4 percent of overall energy use. Rickover, as well as White, will likely be proven right about mankind's inevitable shift from fossil fuels to a final, ultimate state of nuclear power.

Chapter 6

The Steady-State Economy

One feature that will characterize a nuclear economy as different from the fossil fuel-based civilization is that it will ultimately be steady state in nature rather than growth based. For millennia, humans have gotten used to ever increasing supplies of energy. Albeit slowly, human population and per capita energy consumption have risen as humans transitioned up the lower rungs of White's ladder. Then, with the start of the fossil fuel age, energy consumption and human population skyrocketed. As Rickover remarked,

> It is an awesome thing to contemplate a graph of world population growth from prehistoric times . . . in the 8,000 years from the beginning of history to the year 2000 A.D. world population will have grown from 10 million to 4 billion, with 90% of that growth taking place during the last 5% of that period, in 400 years. It took the first 3,000 years of recorded history to accomplish the first doubling of population, 100 years for the last doubling, but the next doubling will require only 50 years. Calculations give us the astonishing estimate that one out of every 20 human beings born into this world is alive today.

Rickover remarked that world population would reach four billion by 2000, but he was off a little—it ended up at six billion, and is 6.8 billion as of 2009. With the transition to the fossil fuel age, humans were able to raise their standard of living while at the same time greatly increasing their numbers. In the past, a rapid rise in population would have lessened the available resource base and led to a decline in standard of living. In a nuclear economy, world population will hopefully reach a certain height and stop growing, which will require changing how the modern banking system functions and the habits of most people. Hopefully, world population will stabilize at around ten billion, and global energy consumption, if used efficiently, will stabilize at around thirty to forty terawatts—roughly double today's energy use rate.

Einstein once remarked that "the most powerful force in the universe is compound interest." Few realize just how powerful exponential growth can be. Since the mid-1800s, fossil fuel consumption grew in the United States by about 7 percent per year, doubling every ten years. This is "the law of seventy-two." Simply divide the percentage rate of growth by seventy-two to get its doubling time. For instance, even at just one percent per year growth, the population of the planet will be over thirteen billion in seventy-two years, then twenty-six billion in 144 years, and fifty-two billion in 216 years. One percent per year growth might not sound like much, but any exponential growth can quickly pile up. The United States was using a million barrels of oil a day in 1920 and three million a day by 1950; today the United States uses nineteen million barrels of oil a day. All the fossil fuels used in the entire nineteenth century wouldn't last more than a couple years at today's consumption rates. Wall Street still expects growth of at least two percent per year. At such rates, oil consumption would double every thirty-six years. In 108 years, oil consumption would be eight times as high, and in 144 years, sixteen times as high, at just two percent growth per year! Clearly, something must change, for eventually highly developed nations must stop growing. We live on a sphere, and the Earth can only hold so many people at a high standard of living.

To get a sense of the power of exponential growth, take a chessboard and put a single grain of rice on the first square, two on the second, four on the third, etc. By the time you cover the sixty-four squared chessboard, you would have used (2^{64}) grains of rice, which is eighteen quadrillion grains of rice, or about 461 billion tons. To put that into perspective, the entire 2008 global rice harvest was about 665 million tons, or 0.14 percent as much. This classic example is said to have bankrupted an ancient king who promised such a chessboard of rice but couldn't come close to paying it. Understanding the power of exponential growth reveals why energy use keeps rising rapidly, even as economies around the world make great improvements in energy efficiency.

All organisms possess the ability of exponential growth, and are biologically programmed to reproduce. Why, then, isn't the world completely covered in organisms? The amount of energy available to them in their ecosystem limits their growth. The amount of energy available must change for growth to rapidly occur, which has happened before with startling effects. In 1859, twenty-four European rabbits were released into the wild in Australia, to provide game to hunt. A decade later, rabbits were so numerous that millions were killed each year without any major stress on the total population. Such growth wouldn't have been possible in Europe, since the number of rabbits had already reached a balance with the amount of food resources available. When introduced into a new habitat, growth was exponential until a new balance had been reached and the continent was filled up with rabbits. A new carrying capacity, or supportable population limit, had been reached. In turn, if a population's energy base is diminished, its environment's carrying capacity shrinks and population declines.

Humans have greatly increased their carrying capacity by clearing inedible forest from fertile lands and replacing it with edible domesticated crops. With the advent of fossil fuel powered tractors and fertilizers, further growth became possible. As fossil fuels deplete, a massive new source of energy will be needed to maintain the current human carrying capacity, let alone provide food energy for even more population growth. Charles Darwin developed his theory of evolution in part based upon the teachings of Thomas Malthus, who in 1798 demonstrated that if human population growth continued to grow exponentially, it would be checked by famine unless some technological means of greatly increasing food production was invented. While England at the time was just beginning to tap its best coal seams using Watt's steam engine technology, and industrialized agriculture would follow, Darwin realized that energy limits were the basis of evolution since only those organisms best able to obtain food energy would come to occupy their environment's carrying capacity. As Darwin put it, there wouldn't be standing room on the Earth given the power of exponential growth if not for these limits. Darwin's grandson Charles Galton Darwin would later write a book called "The Next Million Years" claiming that the only possible way to support a large human population indefinitely was controlled nuclear fusion in reactors on Earth. The fission breeder is now known to provide another.

As fossil fuels reach their limits and exponential growth in demand continues only atomic energy will be able to replace them. In July 2008, the "China Economic Blog" projected that the number of cars in the world will increase by 2.3 billion from 2005 to 2050, with 1.9 billion of the increase in developing nations (21). There are 800 million cars in the world today, so clearly those 2.3 billion cars are not going to be fossil-fuel powered, for even if there was enough oil in the world to power that many vehicles, the global warming costs would be catastrophic. Likely, they will be electric, and likely, they will be nuclear-powered.

Since banks loan out far more dollars than they have on hand, growth is assumed, which will have to change in the future. As we make the transition to nuclear power, and eventually the entire world reaches a high level of growth, a steady-state society will have to emerge based upon the physical limits of the planet to only hold around ten billion inhabitants. In 1974, during the first oil crisis the United States faced, M. King Hubbert testified before the House of Representatives hearing on National Energy Conservation Policy. He remarked,

> I was in New York in the '30s. I had a box seat at the depression. I can assure you it was a very educational experience. We shut down the country because of monetary reasons. We had manpower and abundant raw materials. Yet we shut the country down. We are doing the same kind of thing now but with a different material outlook. We are not in the position we were in 1929-30 with regard to the future. Then the physical system was ready to roll. This time it is not. We are in a crisis in the evolution of human society. It's unique to both human and geologic history. It's never happened before and it can't possibly happen again. You can use oil only once. A non-catastrophic solution is impossible unless society is made stable. This means abandoning two axioms of our own culture, the work ethic and the idea that growth is the normal state of life.

The United States was using ten million barrels of oil a day in 1974, and today about twice that amount is used. The United States was able to continue to grow after the oil peak of 1970 due to increased oil imports, the massive oil field Prudhoe Bay in Alaska, and improved efficiency in the use of oil resources. However, efficiency has diminishing returns for as technology approaches maximum efficiency there is less room to squeeze every last drop of energy out of an energy source. And as the world peaks there will no longer be room to continue importing. Even worse, countries export their surplus, meaning that the United States could be cut off from oil imports relatively quickly. For example, if a country exports a third of its oil production, then if its production goes down by a third, its exports will be totally halted since its surplus has been diminished. The United States is today even more oil dependent than in the '70s, and even more vulnerable. We don't have the easy-to-access natural resources we had during the depression era of the 1930s Hubbert refers to. The time for nuclear power is now, and as rapidly as possible. But even as a nuclear economy develops, it is prudent to keep in mind that eventually the era of growth that has existed since prehistoric times will end. Growth must continue today, but centuries from now the world will inevitably have transitioned to a steady-state, sustainable society powered by atomic energy. The Earth just isn't big enough to hold unlimited exponential growth.

PART III

The Alternatives

To be blunt, there are fossil fuels and there is nuclear.

—Dr. Charles E. Till, Co-developer of the Integral Fast Reactor, an inherently safe nuclear reactor with a closed fuel cycle, September 2005

Chapter 7

Electricity

Electrification will be a vital component of the future energy mix. Solving the problems of the global economy, peak oil and climate change will follow electricity generation closely. Electricity is an extremely energy dense, flexible and efficient form of energy. Because of its unique properties, electricity has been used in ways that would not have been possible with other forms of energy. The field of electronics—from the internet, television, radio, videogames, computers, and scores of other uses based on semi-conducting microchips—have all been possible only because of electricity's unique properties. Communications, entertainment, and information processing were revolutionized in the twentieth century based on growth in electricity use.

During the twentieth century, electricity consumption increased faster than overall energy consumption. When it comes to economic growth, the most important commodity has been electricity. As Peter Huber and Mark Mills put it in their 2005 book "The Bottomless Well," "Economic growth marches hand in hand with increased consumption of electricity—always, everywhere, without significant exception in the annals of modern industrial history." U.S. consumption of overall energy rose ten-fold during the twentieth century; electricity consumption rose thirty-fold. Electricity will likely play an even bigger role in the future than it does today.

Since Thomas Edison lit his factory on electricity for the first time in 1879, entirely new industries—such as aluminum production—have arisen that are entirely dependent on electricity consumption. Even transportation has been revolutionized by electricity; in 1850, the edge of the city of Boston lay a mere two miles from the city center, but by 1900 electrified mass transit had allowed the city perimeter to spread ten miles from the business district.

Most of the conveniences associated with modernity are run on electricity, and even a small increase in electricity consumption can profoundly improve the standard of living of the poor. Two billion people worldwide still rely upon unsustainable biomass combustion as their principle source of energy, including dung, straw, and slash-and-burn forest destruction. An "Asian brown cloud" now extends from India to China and Southeast Asia as a result of burning this biomass and coal. Simple cooking and lighting technology running on clean electricity could largely phase out this air pollution. This would greatly reduce asthma-causing emissions and global warming gases released by burning animal manure and switch grass for home heating and cooking in impoverished nations that lack wood, such as India. Much of the air pollution around the world could likewise be eliminated through electrification.

Unfortunately, however, electricity doesn't come from the wall outlet. It needs to be generated from a primary source. As such, electricity is only a *carrier* of energy—a way of moving energy from point a to point b. This is a concept a surprisingly large number of people fail to grasp. The 2006 documentary "Who Killed the Electric Car?" argued that more electric cars wouldn't cause more air pollution even if the electricity comes from coal. Of course it would, since coal plants are about 40 percent efficient—that is, of all the energy in the coal they burn, 40 percent is converted into electric energy and the rest is lost to inefficiencies of the generation process. By the time the electricity reaches the car, due to line loss traveling through the wire and inefficiencies of charging the battery, only about 30 percent of the energy of the coal is used. In contrast, the internal combustion engines of new cars also approach 30 percent efficiency. Since coal is a much dirtier fuel than gasoline, if you get the electricity from coal, you don't come out farther ahead, you come out behind.

In his excellent book "Hot, Flat, and Crowded," Thomas Friedman argues that we need electrons that are abundant, clean, cheap, and reliable. As such, *the success of electricity is entirely dependant upon the primary source of energy used to generate those electrons.* If the primary energy source is either not abundant, or not clean, or not cheap, or not reliable, we're in for some major hurt. Only one energy source is capable of providing all four categories: nuclear fission. Electrifying the world using nuclear power plants can solve the world's problems, and become the "new oil."

In the future, electricity will be used for home heating, illumination, cooking, appliances, industrial processes, and even transportation. Factories that currently use coal to make steel can switch to electricity. High-speed electric trains could go a long way in reducing the need for trucks and airplanes, and electric vehicles could replace the internal combustion engine. In 2011, Tesla Motors will be releasing a fully electric sedan, the Model-S, which will get 160 miles to the charge and recharge in only forty-five minutes (22). The car is too expensive at $49,900, but as technology improves the cost will likely come down over time. At current electricity prices, the cost of recharging the car will be less than $4, extremely cheap compared to what it costs to fill-up at the gas pump today. In 2010 Nissan will be releasing a fully electric vehicle, the Nissan Leaf, capable of a 100-mile range and possibly priced as low as $25,000. Hopefully, the electricity just won't come from fossil fuels in the future, which is still how the world gets most of its electricity, including the United States.

CHAPTER 8

Solar

This one is a favorite for a lot of people, and it is easy to see why. The Earth is continually bathed in 165,000 terawatts of clean energy from the sun, which comes out to about 1.35 kilowatts of energy for every square meter of surface upon which it directly shines. Enough energy hits the Earth from the sun in one hour to power the 15 terawatts modern civilization currently consumes in a year. All other "renewables" are tied to solar, except geothermal, which is tied to nuclear heat. One passing glance at a solar panel, and you're dreaming of a clean energy utopia where you just scale that panel up a trillion times and have all the clean energy the world will ever need. There are two main ways to use solar energy—passive solar, or the direct use of solar energy for heat and light, and solar panels, which collect solar energy to generate electricity.

Passive solar is great. Just open up a window and you're bathed in clean energy from the sun. The ancient Greeks and Chinese built their homes to face south, which provides shade in the summer when the sun is directly overhead at noon, but allows sunlight to pour in for home heating in the winter when the sun is at a lower angle at midday. Buildings were once designed to let in more passive solar light during the day, as well as passive wind, which is the result of the sun's differential heating of the Earth. Architecture of the antebellum South commonly utilized high ceilings to allow passing breezes to supply cooling in an uncomfortable climate. Let in a cool breeze and you're saving money on air conditioning. Six percent of electricity used in the United States is the result of clothes driers. Replacing clothes driers with clotheslines would use sun and wind to dry clothing rather than electricity from the grid. Passive solar is cheap, efficient, and has been used since ancient times.

Solar panels, on the other hand, are another matter entirely. The type of solar panel most people are familiar with is the photovoltaic panel, which uses exotic materials like cadmium and silicon to generate electricity directly upon contact with light. Even in an environment favorable to solar like the southwestern United States, providing an amount of power equal to a typical household's energy consumption, about a hundred kilowatt-hours per day, would require a few thousand square feet of solar panels on one's rooftop. Most rooftops simply aren't anywhere near large enough to provide this much power. This assumes electric space heating, cooking and an electric car in the garage. The cost of such an installation would be $40,000-$100,000. It would last about twenty to thirty years until the solar panels wore out. The cost would be about $8,000 per kilowatt of capacity, or about twice the cost of a new nuclear plant's $4,000 per kilowatt of capacity.

Capacity *factor*, in contrast, is the real cost concern. Capacity factor is the actual power output relative to the theoretical maximum power output. Solar panels have a low capacity factor since the sun isn't always overhead. The average capacity factor for photovoltaic solar in the United States is 14 percent. The Achilles heel of solar panels is that they only work with efficiency when the sun is directly overhead. It doesn't work at night, or anywhere near as well in the winter. Mornings and evenings are also troublesome, as are clouds.

So, on average, solar panels only produce about 14 percent of what their capacity actually is. Nuclear plants, in contrast, produce reliable power 24/7. This reflects a high return on investment, for solar a poor return on investment. Even though the capacity is twice as expensive, since solar only works 14 percent of the time, it is vastly more expensive. The average cost of solar is twenty to fifty cents per kilowatt-hour averaged over a day/night cycle. For nuclear, the average cost is under two cents per kilowatt-hour for existing U.S. nuclear plants.

Another concern with a low capacity factor is that it necessitates energy storage. Uranium or fossil fuels are stored energy, and as such can pump electrons into the grid day/night, summer/winter, exactly as demand dictates. This is called base-load. Solar is more accustomed to peak-load, or certain points during the day that demand for electricity peaks near noon and solar works with efficiency. *As such, a regime based on solar, no matter how much land was covered in solar panels, would require energy storage.* To do this using lead acid storage batteries would require a battery the size of a large room for every household and cost $20,000-$40,000. The lead acid storage batteries would need to be replaced every two to three years, and at special landfills to avoid toxic chemicals from leaking into the water table. As such, the cost of new lead acid storage batteries every few years and new solar panels every few decades would be far too great, and reduce a population to poverty.

A single eight-pound gallon of gasoline recently priced at a couple dollars contains 125,000 BTUs of thermal energy, the energy equivalent of 500 hours of human labor. To store this much energy using today's most cost effective battery technology would require one ton of lead acid storage batteries (11). Most people are shocked to find these facts out, but it makes sense if you think about it. A single gallon of gas can take a Prius fifty miles. How long would it take someone to push a Prius fifty miles? A long time, but it would be possible if he or she spent a few hours a day for a few months. What allows fossil fuels to be so energy dense is the force of magnetism holding the electron arrangement of hydrocarbon molecules together. It is unlikely a battery will ever approach the energy density of fossil fuels. For instance, even advanced lithium-ion batteries of the most expensive electric cars contain as much energy as two gallons of gasoline. Their mileage is only good because they use their precious store of energy extremely efficiently.

Moore's law, formulated in the 1960s, states that the ability of silicon microchips to store information doubles every eighteen months. This has led some to think all technologies will progress exponentially. Silicon-based solar panels, in contrast, don't store information—they harvest energy. Nothing can change the concentration of the weak influx of solar energy to the Earth, or the amount of land required to harvest that energy. In the continental United States, this influx averages 0.63 kilowatts per square meter at mid-day in June. Averaged over a day/night, summer/winter cycle, and taking into account solar panel inefficiencies, solar panels can only capture about 0.03 kilowatts per square meter. In contrast, the state of New York averages 0.055 kilowatts per square meter of energy consumption over its land area. Covering every rooftop would be a very tiny contribution, and even covering the entire state would only be half the power consumed. This would of course be impossible, given that land is already occupied for roads, farmland, etc. and space needed to store all that energy.

There is just no getting around the fact that untenably large areas of land have to be covered to capture sufficient quantities of solar energy. While solar panels are becoming thinner, which is reducing costs, this can also lower the efficiency and durability of the panels. As such, there have been no significant drops in solar panel costs since the late 1980s, although before that there were significant drops in cost. Revolutionary materials advances in nanotechnology will be needed for a significant change in the efficiency and cost of solar panels, which may never occur. Scientists only have the periodic table of the elements to work with, after all.

It is hard to imagine making all the batteries and solar panels that would be needed using solar energy. Presently, the carbon footprint for PV solar is sixty grams of CO_2 per kilowatt-hour. For coal it is 900 grams, for nuclear six grams. Solar panels are manufactured under high temperatures using fossil fuel energy, as well as associated storage batteries. The manufacture of solar panels requires highly toxic heavy metals, gases, and solvents that are carcinogenic. Brookhaven National Laboratory's department of environmental sciences has amassed 150 studies documenting the toxic poisons, flammable substances and hazardous chemicals used to manufacture solar panels, which are vented into the environment from solar plant manufacturing facilities. After the twenty to thirty years a solar panel lasts, it must be properly disposed of in special toxic waste dumps. Heavy metals like lead and cadmium could vaporize into the atmosphere if put through a municipal waste incinerator. If panels are dumped into a municipal waste landfill, heavy metals like arsenic could contaminate the water table. Compare this to the statistic that a lifetime of electricity could be provided to a family of four by a piece of uranium the size of a golf ball. The amount of waste from nuclear—some of which mostly decays away within a few centuries—looks pretty good compared with solar pollution, which can last indefinitely and isn't always sequestered.

In Spring 2008, the U.S. Energy Information Administration determined that solar was the most heavily subsidized energy source at $24.34 per megawatt-hour (23). Wind came in at $23.37, nuclear at $1.59, and coal at forty-four cents. Solar would cease to exist if not for heavy tax breaks and governments subsidies, and a non-fossil fuel based economy would be nowhere near affluent enough to afford such generous contributions. To truly move away from fossil fuels, energy sources will need to stand on their own.

Imagine a family who lives in the city decides to move to the suburbs to have a solar-powered lifestyle. Now they live in a larger home that requires more heating in winter, cooling in summer, and insulation, and took more energy to build, whereas in their apartment—with no roof for solar panels—heating and other forms of energy could come right off the grid. No car would even be needed in a downtown urban environment, and all those freeways linking the suburbs to the city wouldn't need to be built. Now the family has to spend $100,000 on solar panels and batteries, which were polluting to manufacture. They cut down the CO_2-sequestering trees providing shade in their yard so the solar panels have more light, then have to use more solar energy for air conditioning. On a hot night, their batteries deplete, and they need to go without power for air conditioning when in fact if they just kept their shady trees their house wouldn't have gotten as hot anyway.

Millions already live in apartments with no roof for solar panels, and given how much lower the energy consumption is for city-dwellers, we should encourage them to remain in their present living arrangements. This is all of course theoretical; most houses wouldn't even have a roof big enough for thousands of square feet of solar panels. When it comes down to it, in the future, electricity has to come off a grid, not from the rooftop. This will require reliable base-load generation from centralized power plants.

More promising is the outlook for concentrated solar power. Concentrated solar uses mirrored arrays or parabolic troughs to reflect the sun's light into thermos tubes running through heat exchangers that transfer the sun's energy to a steam generator to produce electricity for the grid. The process requires fewer exotic materials for plant construction, and since the mirrored arrays are electrically wired to track the sun's movement, their capacity factor is often as high as 2,000 hours per year, or around 20 percent.

Another technology, solar thermal towers, use their reflector arrays to heat molten salts or compressed steam, which through a heat exchanger can drive steam turbines. The molten salt or compressed steam heats up slowly and cools down slowly, smoothing out sudden drops in power from a cloud passing overhead, making it easier for utilities to get more than one percent of their power from solar without risk of a sudden power surge. As the molten salt or steam cools, it can produce diminishing quantities of power a few hours into the night. This should not be confused with base-load or energy storage; molten salt simply smooths out power surges due to clouds passing overhead. In the desert, it works 300 days a year.

In 1999, about 900,000 gallons of heat exchanger fluid, called therminol, caught fire at a concentrated solar plant in Barstow, California in the Mojave Desert. The Solar Electrical Generating System (SEGS-II), built in 1987, consisted of 100 acres of forty-foot high parabolic mirrors that rotate to follow the sun. The fire destroyed most of the facility, and required thirteen fire engine companies and 1500 gallons of foam to be put out. Toxic fumes were released over a half a square mile of land, which had to be evacuated; a no-fly zone was later setup over the facility due to emissions of toxic fumes. At the time, this single plant was 90 percent of global solar generation.

The website "Repower America" claims it is possible to power the country with renewable energy in ten years. According to their site, "a parcel of land fewer than 100 miles on a side in the Southwest could theoretically supply 100 percent of America's electricity needs." Mile-on-a-side terminology confuses the public and leads to misconceptions. This means

100 by 100 miles, or 10,000 square miles, could theoretically power the country. To put that in perspective, there are now only ten square miles of solar panels in the entire world with an output of less than one nuclear power plant. So, scaling something like that up would be a truly unprecedented industrial undertaking.

A January 2008 "Scientific American" article titled "The Solar Grand Plan" calculates what it would take to get solar up and running in this country (24). The article, which is very pro-solar, calculates 30,000 square miles of concentrated solar installations would be needed to provide just 69 percent of U.S. electricity by 2050, or 35 percent overall energy. To put this into perspective, it took sixteen months to build the most famous CSP plant in the country, Nevada Solar One, a seventy-five MW plant with 760 parabolic troughs. The plant covers 300 acres with solar installation. To get 30,000 square miles by 2050, we would need two square miles a day. If it took sixteen months to build 300 acres, at the same rate it would take 5.6 years to build two square miles. So basically, *we would need to shrink a 5.6 year construction time down to a day, and do that every day until 2050, to get just 35 percent of our energy from solar.* Not too promising. And never mind the fact that the lifespan of these plants is itself only a few decades. Such vast areas of land need to be covered in technology that solar just isn't tenable. It took France twenty years to make nuclear power its primary source of energy. Solar may never provide more than a small contribution.

The "Solar Grand Plan" goes on to say that solar energy can be stored using giant underground caverns of compressed air. Electricity from solar plants would be used to pump air into underground caverns, aquifers, abandoned mines, and depleted natural gas wells. The process, the article concedes, would still partially rely on natural gas co-generation. The whole point is that we're supposed to be getting off fossil fuels, and building natural gas plants and compressed air plants on a grand scale would make this project even more untenable. We don't even have anywhere near enough of the right geologic formations to fill with compressed air to make such a storage plan useful on such a grand scale.

To make matters worse, solar mirrored-arrays get dusty or become covered in bird poop and need to be washed every ten to twenty days. Where are you going to get the manpower and water to wash 30,000 square miles every ten to twenty days in the desert? Are we going to divert entire river systems into the middle of the desert to clean these panels? The numbers for solar just don't add up. The energy is there, but it is so diffuse and intermittent that capturing, storing, transporting, and harnessing work from the energy is a truly Herculean task. Maybe in the distant future miracles in materials technology will allow for super cheap and efficient solar panels to charge super high energy density batteries. Advances in nanotechnology may yield a paint-like solar material that can be sprayed across the desert quickly and not wear away or need to be cleaned. Sadly, such technology still only exists in science fiction and today's technology is simply nowhere near where it needs to be for solar to contribute much more than the 0.1 percent of our energy it currently provides.

Chapter 9

Wind

Like solar, wind power is diffuse and intermittent, so harnessing and storing wind energy has thus far proven to be technically very challenging. The "Repower America" website claims that "enough wind blows through the Midwest each day to supply all of America's electricity." In practice, capturing all of that energy is another matter entirely. Windmills range from about 1.5 to 5 megawatts capacity apiece, and are around 300 to 400 feet high. That's about the size of a football field if laid down on its side, for a smaller turbine. Wind turbines can only extract a tiny portion of the energy of the wind around them—the portion that hits the turbines. If the power density of wind facing a turbine is 300 watts per square meter, only about four watts per square meter is extracted. Wind turbines also cannot be lined directly behind each other, but must be spaced apart so as not to block one another's wind. As such, it takes 200 square miles of wind farm to equal the output of a single one gigawatt nuclear power plant, which would cover a third of a square mile. As a result, *wind doesn't scale.* There just aren't enough places with enough wind. The best spots are the California coast and Midwest, which just aren't big enough.

That said, there is still plenty of room to expand wind beyond the tiny 0.3 percent of our energy it provides today. Doing so, however, will likely never result in more than a few percent of our energy coming from wind. The Achilles heel of wind, like solar, is that there is no good storage mechanism. Sporadic and unreliable power is unacceptable. Civilization is heavily dependent upon continuous supplies of electricity, and even short blackouts could be catastrophic. On August 14, 2003, a blackout was triggered throughout most of the Northeast by downed power lines. Social and economic chaos ensued. Aircraft were grounded, trains halted and road traffic was at a standstill, due to lack of traffic lights and fuel. Water supplies were severely disrupted, as were telecommunications, while buildings had to be evacuated due to a lack of fire detection and suppression systems. Living in the dark during blackouts is a major disruption, and fires can start when people resort to candles. Major technical civilizations are completely dependent upon continuous supplies of their lifeblood, cheap and reliable electricity. Energy storage is a must.

In the case of wind, pumped-hydro storage has sometimes been proposed. This technology means flooding a geologic mountainous formation with water, pumping the water uphill when there is excess capacity, then letting it flow back downhill to generate power at other times of the day when the wind doesn't blow. The problem is that this not only makes wind even more expensive than it already is, but this technology is also very geographically limited. There just aren't very many spots with the water or land formations to build artificial hydro dams. As such, very little wind is stored worldwide.

Even without storage, wind has a cost of $3,000 per kilowatt installation, coming in a little cheaper than the $4,000 per kilowatt of western nuclear plants (meaning a large 16,000 megawatt nuclear plant that could light up a city the size of Seattle would cost about $6 billion). However, the average capacity factor of wind in the United States is 27 percent. So, wind is many times more expensive than nuclear, even if you don't store electricity. This reflects a poor return on investment. Wind also has high repair costs. Turbines frequently snap in the wind and are struck by lightning, and ice can lower efficiency by over 30 percent until removed. In fact, on Aug. 25, 2007, a man near Wasco, Oregon was killed by a wind turbine that snapped in the wind (25). That's one more death than in Three Mile Island. Turbines often last only fifteen to twenty years, whereas a nuclear plant, once built, lasts sixty years. As such, wind is very expensive indeed. If something only lasts a quarter as long, and only produces power a quarter of the time, that's a cost increase by a factor of sixteen.

In 2008, an independent assessment of Britain's wind industry found that wind is not only far more expensive and unreliable than once thought, but also requires large quantities of natural gas (26). As it turns out, *the high efficiency base-load power plant was not designed for load cycling*, or the fluctuating of its power output to balance wind's sporadic and unreliable power output (26). As little as a 20 percent decrease in wind speed can halve a turbine's power output, and the wind never blows steadily. Only specially designed gas-fired combustion engines or hydro dams are able to cope with these rapid fluctuations. This is bad news for wind, since coal and nuclear happen to be the cheapest, most abundant ways to generate reliable electricity.

Britain, which imports most of its gas from Russia, found that wind would increase the nation's reliance on expensive gas imports. To make matters worse, load cycling gas plants are less energy efficient, supplementing over consumption of fossil fuels and offsetting wind's contribution. They also suffer from more wear and tear and have higher maintenance costs than traditional base-load plants. Unless hydropower can be used to back up the wind, why bother?

The claim is often made that Denmark gets 20 percent of its electricity from wind, but 84 percent of that power is actually exported to Norway where hydropower serves as load cycling (27). Denmark has also maxed out its wind resources, and has placed a moratorium on wind expansion. Niels Gram of the Danish Federation of Industries has remarked "In green terms windmills are a mistake and make economically no sense. Many of us thought wind was the 100 percent solution for the future but we were wrong. In fact, taking all energy needs into account it is only a 3 percent solution." And yet, the EU still has quotas in place to get 20 percent of its electricity from renewables—primarily wind—by 2020. How in the world will they balance that 20 percent? Massive gas imports from Russia and inefficient load cycling plants?

Germany is the world leader in installed wind and photovoltaic solar capacity, and gets six and a half percent of its electricity from wind and one half of one percent from solar. To provide backup, coal plants called "shadow stations" are built to provide electricity for when the wind doesn't blow. Since the shadow stations can't be turned on or off with the flick of a switch, or perform load cycling by quickly fluctuating their power output, they still run when the wind blows, supplementing over consumption of coal. As such wind power is used almost for fashionable purposes and energy mandates rather than for useful energy.

In a 2005 wind industry report, "Wind Report 2005," E.ON Netz, the largest grid operator in Germany, reports "wind energy cannot replace conventional power stations to any significant extent. The more wind power capacity on the grid, the lower the percentage of traditional generation it can replace" (28). When providing about 3 percent of electricity or more, wind requires inefficient load cycling. Above 20 percent, nonexistent complete electricity storage is needed. The wind report also reveals that entire geographic regions may go for up to a week with no wind. This means that "smart grids" that try to send excess power from regions where the wind is blowing to other regions won't work. High barometric air pressure patterns tend to spread across very large continental regions at once, meaning whole groups of countries may go for long periods with no wind. Even worse, this may happen during summer heat waves, when demand for electricity peaks. In a massive heat wave felt in the western United States in late July 2006, high barometric air pressure stifled offshore wind, and California's offshore wind farms contributed nearly nothing.

Without large-scale, long-term power storage, at low-cost, wind will never provide more than a few percent of most geographic regions' electricity. While wind-swept Denmark has accomplished 20 percent wind, Scandinavia as a whole shares that electricity and the overall geographic region derives barely over a few percent of its electricity from wind, and even that is thanks to abundant hydropower resources for load cycling. Now that Germany's high-grade coal deposits are depleting, much of the coal that fuels its "shadow stations" is lignite, a much dirtier form of brown coal with much higher pollution levels than bituminous black coal. Germany has a strong green party, but is failing its emissions targets miserably. In Europe as a whole, greenhouse gas emissions actually rose 1.1 percent during 2008. In contrast, right across the border from Germany, 80 percent nuclear France has Europe's lowest per capita emissions, except for Sweden, which is half nuclear and half hydro. Given the experience in Europe, wind has been blown away.

Chapter 10

Wave and Tidal

The first wave park in the world, the Agucadoura wave park, opened off the coast of Portugal on September 23, 2008. It has a capacity of 2.25 MW, meaning it produces roughly *one percent* as much power as a large coal or nuclear power plant. Connected sections of the plant flex and bend relative to one another as waves run along the structure, and the motion is resisted by hydraulic rams which pump high pressure oil through hydraulic motors which in turn drive electric generators. At La Rance near Cherbourg in France a tidal plant that pumps water uphill at certain beneficial times of the day has supplied commercial power to the grid since 1966, supplementing the mainly nuclear French energy supply.

These technologies remain extremely small players, even smaller than wind and solar. Wave technology is in its infancy, and just beginning to be deployed. However, the problems it will run into will likely be similar to wind and solar. Wave energy is intermittent; it has a capacity factor comparable to wind. Wave parks are expensive to install and can be damaged by storms and saltwater corrosion. Imagine what a powerful storm could do to the electric grid if it takes hundreds of square miles of wave park floating out at sea attached to buoys to equal the power output of a single conventional power station! Tidal also only works at certain times, and is very geographically limited; it only works in certain geographic formations that can be dammed that are near a coastal area.

Chapter 11

Geothermal

Unlike other "renewables," geothermal doesn't derive its energy from the sun, but rather from the stored supernova energy released by uranium and thorium atoms as they decay within the Earth. Since this decay is steady, geothermal power is blessed with a high capacity factor and can run day and night. As such, in some nations geothermal is a large player, providing 26 percent of the electricity consumption of highly volcanic Iceland. However, until recently, geothermal power has been tapped exclusively on the edges of tectonic plates where higher temperature geothermal resources are available. *As such, geothermal will require advances in technology to be significantly expanded beyond its present contribution.*

In the United States, geothermal presently provides 2,687 MW, or about as much power as two nuclear plants. Remaining geothermal sites are geographically limited to the western half of the country, and are located on protected federal parklands and natural wilderness areas that are not open for development. The thermal efficiency of geothermal is low at around 10 percent compared to steam from boilers. By the second law of thermodynamics this relatively low temperature limits the efficiency of heat engines in extracting useful energy during the generation of electricity. Therefore, geothermal has been geographically limited to regions of high volcanic activity with high enough temperatures that are open to development. There are, however, high hopes that geothermal can be expanded in the future beyond its present tiny contribution using binary cycle power plants that may be able to accept temperatures as low as 135°F. This technology is still in its infancy, is unlikely to make geothermal a major player, and would have higher costs and environmental consequences.

Heat extraction for even traditional geothermal power is already expensive, as it necessitates drilling a mile or more underground and requires extensive infrastructure including power lines and cooling water. Drilling so far underground can cause earthquakes, such as when a 3.4 earthquake on the Richter scale and subsequent aftershocks were felt in Basel, Switzerland during geothermal drilling. The process releases groundwater contaminants including carbon dioxide gas, hydrogen sulfide, and mineral rich sludge containing mercury and other toxic heavy metals. Although geothermal remains a tiny player, worldwide it is still larger than wind, solar, wave and tidal combined.

Chapter 12

Hydroelectric

Hydroelectric power involves either simple water wheels that spin by the run of the river, or large dams that can provide base-load generation. The key advantage of large hydro is that water behind a dam is stored energy, like uranium or fossil fuels. As such, the water level behind the dam represents storage capacity that can be run through power-generating turbines faster during the day when power demand is greater, or run through slower at night to allow water levels to build back up. As such, entire cities and even countries around the world with relatively low population densities and suitable river resources can be powered mostly by hydropower, including Norway, Canada, Iceland, and New Zealand. In Norway, hydropower and the imported wind power it compensates for provide 99 percent of electricity.

Hydro is great. The cost of hydro is low at around $1500 per kilowatt of capacity. Hydro also has an excellent capacity factor and relatively low maintenance costs compared to other renewables. Hydro dams have very long life spans, usually hundreds of years. The Hoover Dam may last a thousand years. As such, hydroelectric power remains the cheapest form of power generation in the world. In the United States, hydro dams over fifty years old produce power for just one and a half cents per kilowatt-hour, compared to nuclear and coal at around two cents per kilowatt-hour. Over reliance on hydro, however, can lead to rolling blackouts, as commonly happens today in Brazil during droughts, when water levels behind dams are lower.

Hydroelectric power played an important role in electrifying the United States during the early twentieth century. Around the turn of the twentieth century, conservationists emphasized hydropower as a way of protecting the environment through flood control and shifting away from fossil fuels to more sustainable energy. Theodore Roosevelt was a strong hydropower proponent, stressing it as aiding the "wise use" of natural resources. By the 1920s most city dwellers in the United States had access to electricity, but most rural farmers did not. Hydropower, often located closer to rural areas, was the perfect solution. During the 1930s Franklin Roosevelt's New Deal stressed public works projects including more hydro dams. The Tennessee Valley Authority power company was created to extend power to the rural southeast, and built about forty hydroelectric dams from 1933 to 1944. By 1939, 25 percent of rural households in the United States were electrified, and hydropower was providing a third of total U.S. electricity.

It is no surprise that the developing world is now aggressively building more hydroelectric dams to make use of this relatively benign source of clean, cheap, reliable electricity. Nations include Chile, Brazil, China, and many African nations. China is currently building a massive sixteen GW project, The Three Gorges Dam, along the Yangtze River, which will be the largest in the world. The dam will produce as much power as about ten of the biggest nuclear power plants. 1.3 million people were forced to relocate to allow the dam to flood a vast land area, and in 1975, sixty-two smaller dams in China burst after being damaged by Typhoon Nina, killing 171,000. However, in China over a million die every year due to respiratory illness from coal burning, which is now the #1 cause of death in the nation. Sixteen of the twenty most polluted cities in the world are now located in China. Hydro is benign in comparison.

Recent political stability in Africa has restarted prospects for building an even larger Grand Inga Dam, which will be thirty-nine GW in capacity and be capable of generating as much power as the entire continent of Africa currently uses. The project will take advantage of a natural drop in the Longo River of about 100 meters, which will allow only a modest dam to support a massive reservoir capacity. Without a doubt, hydro has been the most successful of the "renewables" thus far, and is currently the best way of capturing solar energy available. After the sun evaporates moisture against gravity, the tendency of that water to self-concentrate in rivers and back up behind dams makes it vastly easier to harness and store.

In the United States, about 75,000 dams spanning 600,000 miles of river provide about 7 percent of electricity consumption, or about 2.6 percent of overall energy consumption. While the developing world is still expanding hydro, *in the United States and most of the developed world, hydroelectric power cannot be expanded.* The best sites along rivers with the right geologic canyon or mountain formations that can support flooding to form a dam have already all been used. The few possible remaining sites would require either relocating large numbers of people or flooding protected wildlife preserves, and as such aren't likely to be developed. In fact, since 1990, the United States has removed 200 hydro dams, let alone develop more. Globally, hydro provides about 17 percent of electricity and could perhaps be doubled—but by 2050 global energy demand will double. As such, this reliable source of cheap energy will remain a current part of the energy mix but cannot be sufficiently expanded to replace dwindling and polluting fossil energy sources.

Chapter 13

Biomass

Unlike other solar-derived sources, biomass isn't used just for producing electricity, but can be used to produce liquid fuels for transportation, most notably corn ethanol. Since plants remove CO_2 from the atmosphere during photosynthesis, storing it in the form of a durable hydrocarbon, that hydrocarbon energy can be used to power portable machines using standard internal combustion technology without need for an electric battery. However, limitations include the amount of land required to grow such crops and the amount of energy required to fertilize, harvest, refine, and transport those fuels.

The United States currently uses 21 percent of its corn crop to produce 6.5 billion gallons of corn ethanol, or just 1.3 percent of the energy equivalent of our nation's entire oil consumption. In fact, counting soybean diesel, the United States uses 30 percent of its farmland to produce 2 percent of its fuel. *Even if we used all our farmland, we would not produce much more biofuel than we already produce.* Brazil produces biofuel from sugarcane rather than corn or soy, and even if we imported the entire 6.8 billion gallons of mostly sugarcane ethanol produced in Brazil in 2008, it would only be equal to 1.4 percent of the energy content of total U.S. oil consumption (29). Brazil has an economy a tenth the size of the U.S. economy, and much lower fuel consumption levels. Brazil produces enough sugarcane ethanol to run most cars on E25, a gasoline blend that's 25 percent ethanol. However, if Brazilian ethanol production is compared to U.S. consumption levels, it becomes pitifully tiny.

What really invalidates biofuels as a serious energy source is the fact that producing them requires large expenditures of energy and freshwater, particularly for corn ethanol, which also leads to environmental destruction. In fact, *to produce one calorie of food energy in the United States requires ten calories of fossil fuel energy* (7). Feeding Americans is extremely energy intensive, as discussed in the essay "Eating Fossil Fuels," by petroleum geologist Dale Allen Pfeiffer (7). According to Pfeiffer, 400 gallons of oil equivalents are expended annually to feed the average American—more than a gallon a day. To produce food, this includes natural gas used to manufacture nitrogen fertilizer through the energy-intensive Haber-Bosch process (31 percent), diesel for operation of field machinery (19 percent), diesel for transportation of food (16 percent), electricity for irrigation which often still comes from coal or natural gas (13 percent), raising livestock (8 percent), petroleum based pesticides (5 percent), crop drying (5 percent), and miscellaneous (8 percent). Other energy costs include packaging food, refrigerating food and cooking food.

Modern agriculture techniques like nitrogen fertilizer production and tractors revolutionized food production in the twentieth century, based upon increased inputs of fossil fuel energy. This allowed the human carrying capacity of the planet to greatly expand, so our very existence is now completely dependent upon vast fossil energy inputs for industrial agriculture. For instance, when the Haber-Bosch process of nitrogen fertilizer production was invented in 1908, it greatly increased the amount of food energy available since farmers could now intensively exploit all their land every year without depleting the soil of vital nutrients—but the process was very fossil energy intensive. *As oil and gas resources deplete, we will need to find more energy to power industrial agriculture just to feed ourselves, let alone produce fuel.* In fact, it is likely that corn ethanol production requires more energy to produce than it actually provides. If an EROEI is negative, a technology fails to be an energy source. Some argue that this is okay, since much of the energy input is from natural gas used to make the nitrogen fertilizer, and we need liquid fuels to power our vehicles. However, if ethanol is just recycled fossil fuel energy, it ceases to be an energy source. In the future, we will likely have to use electricity from nuclear power plants to produce hydrogen from water via electrolysis to make nitrogen fertilizer.

According to Dr. Tad W. Patzek, professor of civil and environmental engineering at UC Berkeley, and petroleum engineer Dr. Michael J. Economides, to produce enough ethanol to replace one gallon of fossil gasoline, farmers and processors currently consume 1.8 gallons of fossil energy (30). That is, *ethanol consumes about twice as much energy as it produces*. Production costs include 40 percent of the caloric value of the ethanol coming from fossil fuels, mostly for fertilizer production and tractors. Astonishingly, another 58 percent of the equivalent BTU content of the ethanol is used to convert solid corn into a liquid fuel. Another 11 percent of those BTUs are used to transport the liquid fuel to the consumer at the service station. The numbers for ethanol just don't add up. Not only will corn ethanol only increase our dependence on fossil fuels, it will increase global warming gas emissions from fossil fuels burnt to produce it. Fertilizer runoff from ethanol production produces nitrous oxide, a greenhouse gas 310 times as potent as CO_2. In fact, producing one gallon of ethanol requires 2,000 gallons of freshwater, and it takes electricity to run the irrigation systems that provide that water. Ethanol is an environmental disaster.

In 2008, food riots erupted around the world after the poor were no longer able to afford food. According to a January 2008 article, "Why Ethanol Production Will Drive World Food Prices Even Higher in 2008," by Lester R. Brown,

> The world is facing the most severe food price inflation in history as grain and soybean prices climb to all-time highs. Wheat trading on the Chicago Board of Trade on December 17th breached the $10 per bushel level for the first time ever. In mid-January, corn was trading over $5 per bushel, close to its historic high. And on January 11th, soybeans traded at $13.42 per bushel, the highest price ever recorded. All these prices are double those of a year or two ago. As a result, prices of food products made directly from these commodities such as bread, pasta, and tortillas, and those made indirectly, such as pork, poultry, beef, milk, and eggs, are everywhere on the rise. In Mexico, corn meal prices are up 60 percent. In Pakistan, flour prices have doubled. China is facing rampant food price inflation, some of the worst in decades.

When it comes to food vs. fuel, clearly humanity's need to eat is more important than our desire for happy motoring. As more farmland was devoted to corn and soybeans for biofuel production, even wheat prices rose due to less farmland being available for the crop. Even worse, as food prices rose and the United States exported less food to hungry nations, more rainforest was destroyed through slash and burn deforestation to allow food to be grown. Vast tracts of Malaysia and Indonesia have been deforested, as well as South America and Africa. This raises the greenhouse effect even more, since vegetation that once sequestered CO_2 is cleared and replaced by crops that require fertilizers, tractors, and have their CO_2 content released back into the atmosphere upon consumption. In effect, humans are destroying their rainforests so that they can burn ever increasing volumes of fossil fuels ever more inefficiently. If we never even made the ethanol in the first place, we could drive diesel cars rather than power the tractors, or compressed natural gas vehicles rather than use gas to make more fertilizer.

By law, many U.S. states are now required to sell E10 gasoline, or gasoline that is 10 percent ethanol. However, a gallon of ethanol only has two-thirds the energy density of gasoline, which lowers mileage. Ethanol can also damage car carburetors and catalytic converters, especially in concentrations over 10 percent. Some states are now considering increasing the required ethanol blend to 15 percent, which would be profoundly destructive to the environment, food prices and automobiles themselves. Ethanol has got to be the worst of the alternatives, and should be abandoned immediately.

Some claim that breakthroughs in advanced next-generation biofuels from cellulose or algae will come to the rescue. However, this hasn't materialized after decades of promises. In 1921, American inventor Thomas Midgley proclaimed to the Society of Automotive Engineers in Indianapolis, "From our cellulose waste products on the farm such as straw, corn stalks, corn cobs and all sorts of material we throw away, we can get, by present known methods, enough alcohol to run our automotive equipment in the United States." In contrast, in April 2009 the cellulosic ethanol company Aventine filed for bankruptcy protection after its stock, trading at $7.86 a share a year earlier, had plummeted to $0.09 (31). Competitor Verenium, also a maker of cellulosic ethanol, had its stock plummet from $4.13 to $0.36 within a year. Turning switch grass, straw and corncobs into motor fuel has proven to be remarkably energy inefficient and expensive. Current practices try to use plant lignin to provide the heat energy necessary for converting cellulose into a liquid fuel, which is unfortunately less than 20 percent of the mass of fast-growing plants. Again, if the EROEI is negative, and a technology consumes more energy than it produces, it fails to be a source of energy.

Algae biofuel isn't faring much better. In fact, 75 percent of companies that have had algal aspirations since the 1980s and 1990s no longer exist. The total amount of algae biofuel the industry has ever produced would amount to only 100 barrels of oil, at a cost of $100 per *gallon* (32). Most previous and current research on algae biofuel has used the algae in a manner similar to its natural state—essentially letting it grow in water with just the naturally occurring inputs of atmospheric sunlight and CO_2. This approach results in pitifully low yields of oil—about one percent by weight. Higher yields—perhaps forty times higher—have been obtained in laboratories by incubating the algae under special fossil-fueled conditions. The algae are often grown on top of coal plants, to provide lots of CO_2 for fast photosynthesis. The whole point is to get off coal, and since the algae uses solar energy to make fuel, there obviously isn't enough sunlight hitting the smokestacks of our coal plants to scale this up. Algae also must be incubated in warm freshwater, and we are already facing freshwater shortages in this country. If the energy to keep the algae constantly warm comes from fossil fuels, this also interferes with net energy. The algae must be fed organic feedstocks, which take energy to make via industrial agriculture. If every step of a process is based upon fossil fuel consumption, the technology ceases to be an energy source. Just like corn ethanol, algae ethanol will likely remain a net energy loser.

Chapter 14

Hydrogen

Since President Bush hyped his "freedom car" initiative in 2003, a random poll of people would likely have found hydrogen fuel cell powered vehicles to be the solution of choice for replacing gasoline-powered internal combustion engines. Given the name "fuel" cell and all the hydrogen hype, it is easy to forgive them. Unfortunately, a careful analysis reveals that the laws of thermodynamics just don't want a hydrogen economy to be possible. In fact, *hydrogen isn't even a fuel, it's a carrier of energy.* Like electricity, hydrogen is simply a way of transporting energy from point a to point b. Free hydrogen only exists in nature in the atmosphere at a concentration of 0.5 parts per million, and isn't useable. There are no hydrogen wells. Hydrogen must instead be split off of compounds that contain hydrogen, which takes energy.

Right now, 90 percent of U.S. hydrogen is produced from natural gas, with 72 percent efficiency (33). In other words, it would be more energy efficient to drive a compressed natural gas vehicle! The whole point is to get off fossil fuels. If hydrogen is made from water, the process most commonly used is electrolysis. To go from hydrogen and oxygen via electrolysis back to water again in a fuel cell via hydrogen combustion is only 25 percent efficient. This is the result of the second law, entropy. Pulling apart water molecules and forcing them back together again loses energy every step of the way. It is, however, possible to produce hydrogen with 50 percent efficiency using thermochemical processes in very high temperature nuclear reactors, where high temperatures and chemical reactions split apart water molecules to produce hydrogen and oxygen using less energy. The inventor of the modern hydrogen fuel cell, Geoffrey Ballard, has called for nuclear power to be used to cleanly and affordably produce hydrogen. A hydrogen economy may just be another name for a nuclear economy in disguise.

What really invalidates hydrogen as a serious contender for transporting and storing our energy, however, is its physical properties. Hydrogen is the lightest element on the periodic table, and is known as the Houdini of elements. Its properties are nothing like natural gas, which is vastly heavier and denser. At room temperature and pressure, it takes 500,000 liters of hydrogen gas to equal the energy content of one barrel of oil, and hydrogen takes up three thousand times more space than gasoline containing an equivalent amount of energy (33). Even worse, hydrogen is extremely reactive with metals, and frequently turns metals brittle through a process known as hydrogen embrittlement, then escapes through tiny pores and cracks in the metal. To make hydrogen a liquid requires either very high levels of pressure or chilling the hydrogen to 423 degrees below 0°F. As a result, hydrogen fuel cells use highly insulated double-walled vacuum-jacketed cryogenic storage tanks that store hydrogen at 10,000 times atmospheric pressure, with special reinforced steel resistant to hydrogen embrittlement. Even with this advanced technology, hydrogen still boils off at 3-4 percent per day (33).

To make matters even worse, hydrogen can combine with ozone to damage the ozone layer. Imagine 800 million hydrogen vehicles replacing the current automotive fleet, with 3-4 percent of their fuel boiling off per day. The ozone layer wouldn't last very long. Hydrogen also combusts spontaneously at room temperature, so having a leaking fuel cell car parked in the garage overnight would likely raise fire insurance rates. Hydrogen was used in the Hindenburg disaster, and can be extremely reactive with other elements and compounds.

A heavy-duty fuel cell engine may have thousands of complex seals, gaskets, and valves. Hydrogen fuel cell vehicle prototypes themselves currently still cost a million dollars apiece, and are estimated to only last two hundred hours until they wear out, or about 12,000 miles of driving. 12,000 miles isn't a very good deal for something that costs $1,000,000. Another reason why hydrogen fuel cells are so expensive is that a typical fuel cell requires twenty to fifty grams of platinum as catalyst. Platinum costs over $1,000 per ounce.

The United States currently has 200,000 miles of natural gas pipelines, which could not be used to transport hydrogen to homes for heating or to hydrogen fueling stations. The hydrogen would quickly make the metal brittle and escape through cracks in the metal. In his book "The Hype About Hydrogen," Joseph Romm, who worked in the DOE during the Clinton administration, estimates that building a hydrogen redistribution system capable of transporting hydrogen would cost 200 trillion dollars (33). According to Romm, "the hyperbolic promotion of hydrogen fuel cell cars as the answer to our energy woes is a scientific and technological wild goose chase, engaged in at our own peril while the global warming clock rapidly runs down." Not only would the supply chain of producing, storing, and transporting hydrogen and its technologically demanding infrastructure be astronomically expensive, but energy intensive as well. Sadly, hydrogen is no more than a distraction from the practical solutions to our energy problems like electrification and nuclear fission that actually hold promise.

There is now even some talk that boron metal may be used as an energy carrier. The metal burns cleanly in pure oxygen, and has four times the energy density of gasoline. The rust it produces as by-product can then have its oxygen driven off at a power plant for recycle. The metal may even be used in a hybrid configuration to supplement the range of battery-electric vehicles. While this technology is in its infancy, and involves extracting pure oxygen from the air, it seems much more promising than hydrogen.

Chapter 15

Energy Efficiency

Efficiency is another favorite for a lot of people, and it is again easy to understand why. Who could oppose getting more from less? Efficiency has been a tenet of energy policies the world over, and should remain so. However, since the dawn of the industrial age when James Watt patented his first steam engine in 1769, *energy efficiency hasn't lowered fossil fuel consumption; it has raised it.* At first this might seem to make no sense at all, but it is a proven economic theory called the Jevons Paradox (20). In 1865, English economist William Stanley Jevons noticed that since James Watt's steam engine was five times as efficient as the previous Newcomen steam engine, it led to robust economic growth, thereby increasing overall energy consumption. People simply built more steam engines and used them more often, which caused the economy to expand rapidly. Jevons wrote, "It is wholly a confusion of ideas to suppose that the economical use of fuels is equivalent to diminished consumption. The very contrary is the truth." In the 144 years since Jevons made that claim, he has yet to be proven wrong on the macroeconomic level.

Jevons published his findings in his book "The Coal Question," which questioned the duration Britain's coal reserves would last. Jevons looked to renewable energy for an alternative, given the fact that efficiency, by lowering the implicit price of energy, would just lead to more economic growth. "The wind," wrote Jevons, "is wholly inapplicable to a system of machine labor, for during a calm season the whole business of the country would be thrown out of gear." He looked at hydro but concluded that "in very few places do we find waterpower free from occasional drought." Considering biomass, he wrote "we cannot revert to timber fuel, for nearly the entire surface of our island would be required to grow timber sufficient for the consumption of the iron manufacture alone." Geothermal was similarly dismissed, for "heat of the Earth, presents an immense store of force, but, being manifested only in the hotspring, the volcano, or the warm mine, it is evidently not available" (20). Of course, in 1865, the discovery of nuclear fission power was seventy-three years in the future. Nobody even knew about it yet.

While it may be true that at the individual, microeconomic level, efficiency can reduce consumption, such as by replacing a worn-out light bulb with an efficient CFL light bulb, on the macroeconomic level overall economic growth tends to swamp and surpass efficiency gains. When population growth and economic growth are taken into account, if there are more light bulbs, or air conditioners, or cars, etc. and they are used more often, overall energy use still tends to rise. Note that Leslie White's law states "culture evolves as the amount of energy harnessed per capita per year is increased, *or as the efficiency of the instrumental means of putting the energy to work is increased*." As economies around the world grow and develop, efficiency just ends up becoming a part of that growth; indeed, efficiency may help encourage growth and make it easier by lowering the implicit price of energy through its increased economical use as a fuel.

Per unit of energy used, the United States today produces more than twice as much GDP as it did in 1950, but total energy consumption in the United States has risen three-fold (20). Our economy is simply six times the size using three times the energy, since we are twice as efficient. In 1970, the average car got thirteen miles to the gallon—about half the mileage of today's cars. And yet, there are 65 percent more cars on the road today, and they are driven 15,000 miles a year on average, since a greater proportion of the population now lives in suburbia. Hence, oil consumption has also doubled, from ten million barrels a day in 1970 to 20.7 mbpd by 2007. The 70-watt Playstation 2 has 10,000 times the processing power of the 1940s ENIAC, using 2,000 times less energy; but a hundred million Playstations worldwide means far more electricity consumption than ENIACs ever used. World energy consumption is expected to nearly double by 2050, even after taking into account great increases in efficiency. So even with maximum efficiency—without which we may be in even worse shape—we will still need even more cheap and reliable energy than we presently consume.

Not everyone is sold on the Jevons Paradox. Amory Lovins of the Rocky Mountain Institute argues in his book "Winning The Oil Endgame" that theoretical super-efficient cars capable of over 100 mpg can allow us to achieve energy independence. The word efficiency appears in his book 549 times (20). When presented with the Jevons Paradox and confronted with the evidence that since the 1970s he has been consistently wrong about past predictions including claims that efficiency will lower energy consumption, Lovins remarked that the Jevons Paradox is "broadly" false (20). According to Lovins, if it were true, inefficient equipment should be mandated.

This isn't going to happen because engineers are constantly trying to wring the maximum amount of work out of a unit of energy, which is a by-product of increased technological development and ingenuity. Automobiles had a thermal efficiency of 5 percent in 1950, whereas today new cars have a thermal efficiency of 30 percent, and for hybrid vehicles and diesel vehicles the thermal efficiency is often above 40 percent. Such vast improvements in efficiency have allowed exponential growth in oil consumption to occur at rates of just 2 percent per year since they were implemented beginning in the 1970s, with oil consumption approximately doubling since then. As such, efficiency became part of the growth, and partially took the place of some of the *increased* consumption.

From 1950 to 1970, oil consumption rose by 7 percent per year, quadrupling over a twenty year time period. Thus, at the very least, efficiency lowers the rate of growth in consumption. However, it has not yet proved capable of countering the power of exponential growth. Modern electric vehicles have efficiency levels approaching 90 percent, and the electricity can come from emissions free energy sources like nuclear. Electric vehicles could therefore take efficiency gains even further than the most efficient internal combustion engines. But by 2050, there may be as many cars in China as there are in the entire world today. Efficiency is a clean way of allowing growth to occur, and we are going to need more efficiency, as well as even more clean and reliable energy, to power larger economies going forward.

Chapter 16

Nuclear Fusion

Fusion has long been a favorite of science fiction authors, due to its promise of unlimited clean energy, and the fact that the technology to perform nuclear fusion will one day likely be in our grasp. Like nuclear fission fuel, nuclear fusion fuel has an energy density about two million times that of fossil fuels by weight. And, like nuclear fission, the fuel exists in seawater in unlimited quantities. Deuterium, an isotope of hydrogen composed of one proton and one neutron, exists in seawater at a concentration of about one part per 5,000—enough to provide twice the world's present energy consumption for about five billion years, or longer than the sun will last.

The sun itself is powered by nuclear fusion, meaning that all other energy sources are in some indirect way the result of fusion. "Renewables" are tied to solar energy, fossil fuels were themselves formed from fossilized photosynthetic organisms, and even uranium and thorium were formed in the massive thermonuclear supernova explosions of dying stars billions of years ago. Fusion, like fission, lies on phase five of White's ladder.

The problem is that in stars massive gravitational forces and temperatures provide extreme conditions allowing the nucleus of hydrogen atoms to become close enough to overcome their natural positive charge repulsion and fuse, thus allowing nuclear fusion to occur and energy to be released. Replicating these conditions on Earth has thus far proved to be technically very challenging. Current techniques use the hydrogen isotopes deuterium and tritium, which are the easiest materials possible to fuse. Tritium is an isotope of hydrogen with two neutrons, and deuterium one neutron. The extra weight of their neutrons helps pull the hydrogen nuclei closer together, allowing fusion to occur under minimal heat and pressure—which is still about two hundred million degrees Fahrenheit.

A five billion Euro fusion plant is now being built in Cadarache, France, named "ITER," or International Thermonuclear Experimental Reactor. The plant is scheduled for completion by 2018, and if all goes as planned, will produce net energy in the form of heat. The EROEI will be around five to ten. It incorporates a half-century of knowledge gained using earlier TOKAMAK magnetic confinement fusion reactors like the Jet, which were unable to produce more energy than they consumed to power the process. ITER will use 0.5 grams of deuterium-tritium mixture as fuel, heated to 200 million degrees as it travels through a donut-shaped toroidal magnetic field, producing 500 MW of power for periods of about eight minutes. Powerful magnets rapidly rotate around the plasma, generating a magnetic field that holds the plasma in place inside a vacuum, since no physical material can support temperatures so great. ITER will operate for twenty years, and if all goes as planned, it *may* be possible to operate commercial fusion plants as early as the 2040s.

While deuterium exists in unlimited quantities, tritium is extremely rare. With a half-life of just 12.4 years, there is no usable tritium on Earth—just trace amounts formed naturally by cosmic rays. As a result, ITER will also have to demonstrate tritium breeding. Since fusion of deuterium and tritium releases its energy in the form of fast neutrons, those neutrons will be slammed into a breeder blanket of lithium lining the walls of the reactor, which will then provide heat for power generation and split lithium into helium and tritium, which can then be used for fusion. As such, the fuel is basically deuterium and lithium. Since lithium must be mined and will be needed for batteries in the future, deuterium-tritium fusion won't have the free fuel supply nuclear fission reprocessing promises. Deuterium-deuterium fusion would require much higher temperatures and pressures than foreseeable technology is capable of. Nevertheless, fusion will likely become a growing part of the energy mix in the second half of the twenty-first century, which is too late to help with the immediate energy problems we face. And even that is under the optimistic scenario that ITER proceeds according to plan. As secretary of energy Steven Chu put it in a speech about energy sources, "I will skip fusion because it will likely skip the twenty-first century."

CHAPTER 17

Methane Hydrates

Deep below the ocean floor, trapped in molecular cages resembling ice, is a huge supply of natural gas that represents at least twice as much hydrocarbon energy as all the conventional fossil fuels in the world. However, since this methane exists in low concentration at about 1 percent pure methane in the mass of the ice, and the ice is a solid form deep below the ocean, economically harnessing it has proven to be technically very challenging. Chevron, testifying to a U.S. Senate Committee in 1998, stated that hydrates occur in low concentration and have no commercial potential. Russian Gazprom likewise came to the conclusion that submarine hydrate extraction is uneconomic. In reality, hydrates may only end up being liberated if global warming melts the frozen methane hydrate ice below the ocean floor and arctic tundra, releasing this potent greenhouse gas. Nevertheless, work on methane hydrate extraction continues in Japan and in India.

Chapter 18

Thermal Depolymerization

In May 2003, "Discover" magazine published an article titled "Anything Into Oil" (34). "This is a solution to three of the biggest problems facing mankind," said Brian Appel, chairman and CEO of Changing World Technologies, the company that built the first industrial size thermal depolymerization plant now up and running near a big turkey processing plant in Missouri. According to Appel, "this process can deal with the world's waste. It can supplement our dwindling supplies of oil. And it can slow down global warming." When reporters interviewing him claimed what he was saying was too good to be true, "everyone says that" was his response. Let the celebration begin.

Private investors chipped in $40 million to develop the process, and the federal government granted more than $12 million to push it along. Thermal depolymerization involves taking organic wastes like turkey guts such as heads and feet that nobody wants to eat, or even raw sewage, and cooking it under high pressure to the point that it breaks down and reforms into energy-dense hydrocarbons with the properties of crude oil. The process mimics the geothermal heat and pressure of the Earth that naturally formed the fossil fuels millions of years ago. And by using organic compounds like turkey guts that are still technically edible, the process requires less heat energy to convert it into oil. Apparently, according to the article, converting turkey guts into oil by the process consumes less energy than is stored in the final oil produced. As the article proudly states, "If a 175 pound man fell in one end, he would come out the other end as 38 pounds of oil, 7 pounds of gas, and 7 pounds of minerals, as well as 123 pounds of sterilized water." Is it time to rejoice?

Not exactly. The process is basically just turkey ethanol. If only we had unlimited "garbage" we could make lots of oil from turkey guts or other animal wastes left over from food production. However, since we only have so much farmland, which captures only so much sunlight to form feedstocks that are then fed to livestock, we can only raise enough livestock to make a pitifully tiny amount of waste body parts. Whereas ethanol from corn is today equal to 1.3 percent of our oil consumption, thermal depolymerization turkey oil is almost too small to mention. Today, the thermal depolymerization plant in Missouri produces 400 barrels of oil a day for a cost of $80 per barrel. Compared to the nineteen million barrels of oil a day the United States uses, that's *less than 1 percent of 1 percent* of our oil consumption. Without all the tractors and nitrogen fertilizer making the food fed to the turkeys, we wouldn't even be able to produce that much. But hey, every little bit helps. Better to turn the turkey guts into oil than simply let it rot. As such, the process makes sense since people eat the good part of the turkey first, unlike corn ethanol, which is grown at a negative EROEI using fossil energy that could have powered our vehicles in the first place.

Chapter 19

Space-Based Solar Arrays and Positron Antimatter

While thermal depolymerization has produced at least some useful energy, and nuclear fusion holds some future promise, space-based solar arrays and positron antimatter are about as hopeless as sci-fi proposals come. In 1973 Peter Glaser patented a technology for transmitting power over long distances as microwave radiation, which eventually led to the idea of using the technology to beam in sunlight from outer space. Between 1978 and 1981, the U.S. congress authorized the DOE and NASA to jointly investigate the possibility of harnessing solar power from outer space. The Satellite Power System Concept Development and Evaluation Program was the end result, which determined that the process was not economical. Most notable was the concern that just transporting material into outer space has an average cost of *$20,000 a pound*. Solar panels built and used right here on Earth, as a result, would be vastly cheaper than solar installations manufactured and transported into outer space. There would also be some EROEI issues with using fossil energy to power space ships that lift solar installations into space at $20,000 a pound. We would not only need untenably large areas of solar panels in outer space just as we would here on Earth, but we would need to build infrastructure to beam the energy back to Earth in the form of microwave radiation and installations to capture it on Earth. If the "Solar Grand Plan" from "Scientific American" isn't possible, this doesn't come close.

The existence of positrons was first postulated in 1928 by Paul Dirac, and discovered in 1932 by Carl D. Anderson. Positrons, a form of antimatter that are the opposite of electrons, convert into pure energy upon contact with matter. As such, *there is no usable antimatter on Earth* since it would instantly convert to energy. The Fermilab Tevatron in Betavia, Illinois uses a four-mile long synchrotron accelerator that slams atoms into one another at such high velocity that antimatter has been created. However, at the rate we currently make antimatter, *it would take a billion years to produce a gram of antimatter*. The process also consumes much, much more energy than it produces, and is only done for research purposes used to study subatomic particles. Nevertheless, pop culture sometimes contemplates the possibility of using this process for energy, like the 2009 blockbuster "Angels & Demons."

CHAPTER 20

Natural Gas

While the fossil fuel companies have created controversy trying to convince the public that coal is "clean," there has been little controversy about just how "natural" gas is. In reality, there is nothing any more "natural" about gas than oil or coal. All three fossil fuels formed from fossilized biomass subjected to millions of years of heat and pressure within the Earth. As a result of its name, natural gas is often perceived by the public and by legislators as a clean alternative to oil or coal. A careful analysis reveals otherwise. Not only is natural gas a major source of global warming gases, it is also tainted with toxic heavy metals and carcinogenic materials.

Natural gas, also known as methane or CH_4, is composed of one carbon atom to four hydrogen atoms. As such, due to this high hydrogen content, it is often noted that gas combustion produces about half the carbon dioxide as coal combustion per unit of energy. While this may be true, *methane gas is itself twenty-three times as potent a heat-trapping greenhouse gas as CO_2*, and its leakage has a very high greenhouse contribution. According to the Society of Chemical Industry's 2004 report, 2 to 4 percent of gas used ultimately escapes during the process of mining, transporting and using gas at power generation plants or for home heating and cooking (35). Due to its potency as a greenhouse gas, methane has a greenhouse contribution rivaling coal. The greatest leakage is usually at the production sites, although when gas flames at homes are turned on and off, a small amount always escapes unburnt, which adds up. Even if just 2 percent of natural gas used each year leaks before burning, it causes over a period of twenty years a peak in global warming comparable to that of burning bituminous coal per unit of energy produced. If 4 percent escapes, the greenhouse effect peaks at more than three times greater than that of burning coal. The good news is that methane breaks down faster in the atmosphere than CO_2, but over a twenty-year period its effects could be disastrous, and be enough to unleash devastating positive feedbacks.

In the December 2008 issue of "Environmental Engineering Science Journal," Italian researchers determined that particulate matter emitted by domestic natural gas burners used for home heating and cooking have been associated with increased mortality due to deposition in the lungs, brain, and circulatory system (36). The particulates measured were found to be in relatively high concentrations in the flame region of home heating burners and stove tops, and were about one to ten nanometers in size. Particulates include the toxic heavy metals mercury and sulfur, as well as the cancer-causing carcinogens benzene and toluene, and the beta emitter radon (37, 38). While most homes in Italy use gas heating and cooking, neighboring France uses electric space heating and cooking powered mostly by emissions-free nuclear power plants. France has the cheapest electric rates in Europe, and is Europe's largest exporter of electricity. Italy, which banned construction of new nuclear plants in 1987, now imports 15 percent of its electricity form France.

Natural gas today supplies 20 percent of U.S. electricity and about 23 percent of overall energy consumption. As such, gas is already an established part of the energy mix. Most of the power plants built in the United States in the 1990s were gas-fired, but now natural gas prices are too volatile for many more to be built. Gas prices, once $1 per thousand cubic feet in the 1990s, reached $13 per thousand cubic feet in 2008. U.S. domestic gas production peaked in 1973, and has been on somewhat of a bumpy plateau ever since.

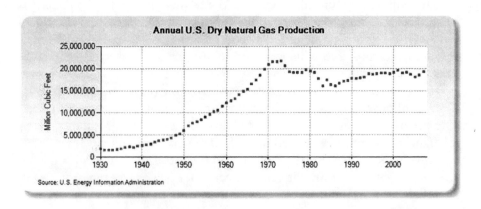

Currently, about 15 percent of U.S. gas is imported from Canada, and a growing amount from overseas in the form of highly explosive liquefied natural gas (LNG). A clear trend has now been established indicating that most large domestic gas deposits have been discovered and used, just like oil. However, gas differs from oil in that production tends to remain high until pressure drops and production of a gas deposit goes off a cliff, rather than forming a bell-shaped curve. This is called the Natural Gas Cliff. Another difference is that gas cannot be quickly imported like oil; it must be liquefied under pressure in cryogenic tankers then re-gasified on shore, which is a slow and expensive process. Tankers and terminals for LNG are expensive, highly explosive and vulnerable to terrorist attacks. We will eventually know the real cost of LNG as a growing portion of our gas will have to be imported. Since oil and gas are already major parts of our energy mix, and their production is in decline, we will struggle to find alternatives to both oil *and gas* in the years ahead.

Chapter 21

Coal

There are four main grades of coal in the world. Ranging in order from highest energy density and desirability to lowest, they include anthracite, bituminous, sub-bituminous, and lignite. The energy density of anthracite and bituminous coal, the black coal most are familiar with, ranges from about twenty-two to thirty million BTUs per ton. Anthracite, the most desirable, also has fewer impurities. Lignite coal is brown in color and has an energy density of just nine to seventeen million BTUs per ton. Sub-bituminous coal, more grayish or mixed in color, is in between. While the United States is often referred to as the "Saudi Arabia of coal" with a 250-year coal supply, the United States actually peaked in the production of anthracite coal, the least polluting, in 1920 at about 100 million tons per day.

Until as late as the 1950s, this desirable type of coal was commonly used for home heating, but has since been replaced with oil, gas and electric. The United States today produces just five million tons of anthracite coal per year. U.S. bituminous coal production peaked in 1990, and has since been supplanted by increasing sub-bituminous and lignite production, mostly from the American West. Whereas West Virginian bituminous coal was once the major source of U.S. coal production, the majority of U.S. coal is now sub-bituminous coal from the massive Powder River Basin of Wyoming, the largest coal seam in the world. About 40 percent of U.S. coal consumption is now sub-bituminous coal from Wyoming, while about half is bituminous from mostly West Virginia and 9 percent lignite from Montana.

Not only does sub-bituminous and lignite coal have an even higher carbon footprint than bituminous coal, more material has to be burnt which means more particulate emissions like mercury or flyash being released into the atmosphere. 60,000 children are born with birth defects each year as a result of mercury poisoning. The mercury gets into the food chain of ocean organisms, and pregnant women are now commonly told to avoid seafood. According to the EPA 24,000 Americans die of respiratory illness due to coal particulate emissions each year (39). At least Wyoming coal is lower in sulfur content, which is responsible for acid rain, for what that's worth.

The United States is estimated to have about 246 billion tons of coal, or about 27 percent of global reserves. Since the United States currently produces 1.06 billion tons of coal per year, this would indeed be enough to last 232 years *at current use rates*. However, the majority of coal left is sub-bituminous and lignite, which not only means higher emissions in the future but a supply that won't last 200 years. Even worse, coal is 49 percent of current electricity production and 22 percent of overall energy consumption. A switch to coal to replace dwindling oil and gas would mean ramping up coal mining to an absolutely unimaginable scale and a coal supply that would likely peak mid-century, even for lignite. The world as a whole has even less coal than the United States and *will likely peak by 2026 in energy from coal and 2034 in overall tonnage* (40). China, the world's largest consumer of coal, has only half the coal of the United States and is using it twice as quickly. China may peak in coal production before 2020. Coal, once $30 per ton in 2000, reached $150 per ton on September 26, 2008.

Coal mining is a terrifyingly destructive process that radically alters the natural environment and kills coal miners. Mountain top removal techniques are turning West Virginia, formerly the mountain state, into a flat region. Politicians sometimes joke that this will make it easier to build more golf courses. The amount of coal that must be mined is so vast, 44 percent of all tonnage transported on U.S. railways is coal. A two gigawatt coal plant requires a 220-car "unit train" of coal cars arriving *every day*. The average American uses twenty pounds of coal *a day*.

Right now, the Achilles heel of coal is carbon dioxide, the heat-trapping gas global warming is attributed to. Unless coal plants are linearly phased out *over the next two decades*, it may be too late to avoid catastrophic climate change "tipping-points" as methane hydrates are liberated. As a result, "clean coal" technology that captures and stores CO_2 may be the only way for more coal plants to be built in the United States, although India and China together are building several dirty pulverized coal plants a week. Clean coal technology involves using a new type of coal plant, the Integrated Gasification Combined Cycle, or IGCC plant. Coal is gasified, scrubbed of pollutants, and burnt as a synthetic gas at about 58 percent efficiency, compared to 40 percent efficiency for standard coal plants. The carbon dioxide given off is then sequestered and injected underground.

That's where the problems start. Not only is the process vastly more expensive at about $29 a ton, but a one GW coal plant also produces several million tons of CO_2 per year (41). A one GW nuclear plant, in contrast, produces just twenty-five tons of spent fuel per year, of which only a single ton is fission product waste that can't be reprocessed for more energy. The United States has 70,000 tons of spent fuel stored away, already "sequestered," and the waste has become a major political conundrum. It will likely be reprocessed in fast reactors for more energy; fast reactors use uranium resources over 100 times as efficiently as today's reactors. A big coal plant, in contrast, can produce 70,000 tons of CO_2 *every four days*. Where are we going to put tens of *billions* of tons of CO_2 from decades of coal consumption? Who will want it under their back yard? Current plans are to force the CO_2 down old oil wells, but they aren't large enough for serious large-scale sequestration. Just capturing and transporting so much CO_2, a lightweight gas, will be a daunting challenge.

China, which gets 70 percent of its energy from coal and in 2008 eclipsed the United States in CO_2 emissions, is now building its second carbon sequestration demonstration plant. Run by Huaneng, the top power producer in China, the plant will capture 10,000 tons of CO_2 per year, or about 0.00016 percent of China's CO_2 emissions (41). According to the EIA, China emitted six billion tons of CO_2 in 2006 and since the 1990s emissions have increased by 8 percent per year, doubling every nine years; a direct correlation with economic growth that has lifted eastern China out of poverty (41). Replacing current pulverized coal plants with expensive IGCC plants the world over, capturing the CO_2, transporting it, and storing it somewhere long term would be a truly massive undertaking.

Since CO_2 remains in the atmosphere for over a millennium, coal companies will have to demonstrate that less than 0.1 percent of CO_2 escapes per year, which is technically impossible. There simply isn't anywhere large enough to put all the CO_2, and no real way to guarantee that geologic uplift, fissures, quakes, and erosion won't result in massive planetary belches of CO_2. In 1986, a large eruption of naturally occurring carbon dioxide from Lake Nyos, Cameroon, killed over 1,700 people over a span of twelve miles in all directions. CO_2 is an asphyxiate, and allowing Americans to live on top of giant underground caverns of CO_2 just waiting to be liberated could be a climate crisis and suffocation crisis in the making. According to Dan Becker, director of the Sierra Club's Global Warming and Energy Program, "There is no such thing as 'clean coal' and there never will be. It's an oxymoron."

Evidence is now emerging that fifty-five million years ago geologic activity released into the air more than a terraton, or a trillion tons, of carbon dioxide. As a consequence, temperatures in the arctic and temperate regions rose eight degrees Celsius, and in tropical regions five degrees Celsius. According to James Lovelock in "The Revenge of Gaia," it was likely over 100,000 years before normality was completely restored. Nearly half that amount of CO_2 has already been put into the atmosphere, and the Earth is already weakened by the amount of CO_2-sequestering forest cleared for farmland. According to Lovelock, if climate change occurs much further, positive feedback loops are unlocked, and the climate changes as much as it did fifty-five million years ago, then "before this century is over, billions of us will die and the few breeding pairs of people that survive will be in the arctic region where the climate remains tolerable." This is coming from one of the world's most revered environmental scientists. Carbon sequestration would likely just be a disaster, patiently waiting to be unearthed, seeping up directly below our feet. Are we willing to take that risk?

While virtually no CO_2 is currently being captured around the world, some coal plants are equipped with scrubbers that capture most of the particulate emissions of coal, such as fly ash, sulfur oxides, mercury, and other heavy metals. It is stored in "slurry dams," giant artificial dams of toxic ash. On December 22, 2008, an ash dike ruptured at an eighty-four-acre solid waste containment area at the Tennessee Valley Authority's Kingston Fossil Plant in Roane County. 1.1 billion gallons of toxic slurry waste was spewed into the Emory and Clinch Rivers, damaging homes and covering over 300 acres in sludge six feet thick. As of June 2009, only 3 percent of the spill had been removed and cleanup will likely cost over a billion dollars. A test of river water near the spill revealed elevated levels of lead, thallium, mercury, cadmium, chromium, nickel, arsenic, and even uranium and thorium. If coal companies can't even ensure that their solid waste won't enter the environment, how likely is large-scale, long-term storage of CO_2 gas?

Not only are slurry dams major environmental risks due to the sheer mass of ash they contain, the ash is also radioactive. Uranium and thorium are such abundant metals in the Earth's crust, there is actually more energy in coal from traces of uranium and thorium than there is in the chemical energy of the coal itself. According to "Scientific American," "the fly ash emitted by a power plant—a by-product from burning coal for electricity—carries into the surrounding environment 100 times more radiation than a nuclear power plant producing the same amount of energy (42)." For those worried about radiation, coal plants are the real offender.

The amount of radiation is still relatively small, much smaller than natural background radiation we are constantly exposed to from cosmic rays and other natural factors, but the fact remains that coal plants expose the public to 100 times more radiation than nuclear plants. Even more ironic, the Clinch River, now flooded in toxic radioactive waste from coal, was in

1983 supposed to be the site of the Clinch River Breeder Reactor, an early type of fast reactor that could be fueled by existing stockpiles of nuclear waste. Nuclear fuels are three and a half million times as energy dense as coal, so nuclear waste produced is small enough that it can be "sequestered" with great care, unlike coal waste. If only contemporary "greens" were better educated in science classes in high school, perhaps the world's environmental crisis never would have materialized in the first place. In the 1950s, the Sierra Club had a slogan "Atoms, Not Dams" proclaiming support for nuclear over hydro, but has now taken a wrong turn and opposes nuclear, like the majority of environmentalists. "Atoms, Not Armageddon" would today be a fitting reappraisal of their past position.

In May 2009, at a corporate climate summit in Denmark, Duke Energy CEO Jim Rogers proclaimed, "I'm betting on nuclear. And I would go a step further and probably say that these two coal plants we build might well be the last two we build until we have a clear picture on carbon capture and storage (43)." Duke energy, the third largest producer of energy from coal in the United States, is building two demonstration IGCC plants, but most of the CO_2 won't be captured or stored. The United States originally had plans for over 150 new pulverized coal plants, most of which were recently scrapped due to global warming concerns, and the high likelihood of a carbon tax, which only expensive and unproven carbon sequestration could avoid by not releasing emissions. In the future, it is likely coal will no longer be cheap, due to high taxes on regular coal plants and the greater cost of IGCC plants, which still haven't demonstrated large-scale, long-term carbon sequestration anyway, and therefore won't avoid carbon taxes. Nevertheless, *the dirty secret of America's energy future is that if we don't build more nuclear power plants, our near-term energy destiny will be handed to coal, climate be damned.*

PART IV

The Nuclear Economy

We must also induce many more young Americans to become metallurgical and nuclear engineers. Else we shall not have the knowledge or the people to build and run the nuclear power plants which ultimately may have to furnish the major part of our energy needs.

—Admiral Hyman G. Rickover,
May 14, 1957, St. Paul, Minnesota

Chapter 22

Phase Five

In 1932, the discovery of the neutron led to scientific speculation that it might be possible to create elements heavier than uranium through a type of nuclear alchemy. A possible Nobel Prize and greater scientific knowledge was now believed to be up for grabs. This led to a scientific race between Ernest Rutherford of Britain, Irene Joliot-Curie of France, Lise Meitner and Otto Hahn of Germany, and Enrico Fermi of Italy to create new elements through bombardment of uranium with neutrons. Since neutrons are neutral in charge, they can easily travel to the nucleus of an atom because they aren't repelled by the nucleus's strong positive charge. This makes possible a fission chain reaction, and the unlocking of untold quantities of energy that have been stored away within the nucleus of the atom for billions of years. Their work would lead to the truly remarkable discovery of nuclear fission—a defining event in human history like earlier inventions of fire or the wheel.

In 1926, at age twenty-four, young physicist Enrico Fermi became a professor of atomic physics at the University of Rome. In 1934, working in the lab of professor Orso Mario Corbino, Fermi's group discovered that neutrons could be slowed by a neutron moderator such as paraffin wax, which has a high concentration of hydrogen and carbon atoms. Small molecular substances such as water, carbon or beryllium slow neutrons and can act as a moderator that regulates a neutron chain reaction. Using paraffin, neutrons were absorbed by uranium, making the atoms unstable which shortly thereafter causes them to break apart. In 1934, Fermi announced he had transformed small amounts of uranium into different elements, and reasoned element ninety-three was created. What Fermi failed to recognize was that nuclear fission had also occurred, creating smaller atoms as well. In 1938, Fermi won the Nobel Prize in physics for his "demonstrations of the existence of new radioactive elements produced by neutron irradiation, and for his related discovery of nuclear reactions brought about by slow neutrons." The discovery of the millennium was barely missed, and would have to wait just a little longer.

In December 1938, Otto Hahn and colleague Fritz Strassman reported that they had detected the element barium after bombarding uranium with neutrons in a manner similar to Fermi. After communicating their results to Meitner, she correctly interpreted the results as being nuclear fission, since the uranium atom apparently split apart into smaller atoms, or fission products. Meitner named the revolutionary new phenomenon "atomic fission." Meitner's nephew Robert Frisch confirmed this experimentally on January 13, 1939. The last phase of White's ladder had finally been reached, and progress was now sustainable. Meitner reasoned that there must have been a huge energy release according to the mass portion of Einstein's equation $E=MC$ squared, and that a self sustained chain reaction may be possible with sufficient neutrons being released. As it turned out, she was correct—and the world would be changed forever.

In 1939, Fermi conducted an experiment at Columbia and discovered significant neutron multiplication in uranium, proving that a chain reaction was possible. The race was on to develop atomic fission. Unfortunately, it was wartime in Europe, and the scientists who discovered atomic fission, many of them Jewish, were fearful that an atom bomb may end up in the wrong hands. As a result, efforts would reluctantly have to be focused on the atomic bomb first, to make sure that it could end the war for the Allies, but producing cheap electricity was the real desired goal.

On August 2, 1939, Einstein wrote a letter to President Roosevelt warning him of this new discovery and its war implications. His letter began:

> In the course of the last four months it has been made probable—through the work of Joliot in France as well as Fermi and Szilard in America—that it may become possible to set up a nuclear chain reaction in a large mass of uranium, by which vast amounts of power and large quantities of new radium-like elements would be generated. Now it appears almost certain that this could be achieved in the immediate future.

The age of atomic fission had begun. Leo Szilard, a colleague of Enrico Fermi, had a friend, Alexander Sachs, deliver the letter to Franklin

The Nuclear Economy

Roosevelt. A Wall Street businessman who visited the president regularly, Sachs made sure to read the letter aloud to Roosevelt—who wasn't familiar with the subject and became weary trying to comprehend. Trying to clarify, Sachs remarked,

> During the Napoleonic Wars, a young American inventor approached the emperor and offered to build a fleet of steamships by which he could cross the English Channel. "Ships without sails?," said Napoleon, and sent him away. That man was Robert Fulton. Historians believe if Napoleon had listened more carefully, nineteenth century European history would have taken a completely different turn.

"What you're saying," Roosevelt replied after pondering the matter, "is that you want to make sure the Nazis don't blow us up." "Precisely," Sachs replied. If Napoleon understood the importance of energy resources in shaping the future, and realized the superior properties of coal that could allow his navy to become vastly more powerful if it transitioned away from wind, European history could have ended up completely different. Just as how coal won wars against wind during the early years of the Industrial Revolution, uranium would now win World War II against nations with only fossil powered equipment. As Admiral Hyman G. Rickover has remarked, when a high-energy society confronts a low-energy society, the advantage always lies with the latter. World War II would be no different. The newly appointed uranium committee would meet the following week, at the Naval Laboratory in Washington, and the Manhattan project would produce an atom bomb by 1945. The amount of mass converted to energy in the bomb dropped over Hiroshima would be 0.6 grams.

Einstein further remarked in his letter, "for the first time in history mankind will be using energy not derived from the sun." Nuclear isotopes store fantastic quantities of energy in their nucleus that have been locked away for billions of years, since before the Earth formed. Only the explosion of a dying star, the supernova, is powerful enough to fuse lighter atoms into very heavy ones, forming elements like uranium and thorium in the process. The Earth later formed from the debris of these exploding stars, complete with large quantities of uranium and thorium atoms. The heat given off by these elements as they decay creates the heat that churns molten iron within the Earth, creating a powerful magnetic field that protects the atmosphere from solar winds. This is why Polar Regions experience northern or southern "lights," which are the charged particles released from the sun accumulating near Polar Regions. Planets like Mars also got uranium and thorium as evidenced by volcanic activity, but likely not as much as Earth. As a result, Mars no longer has a powerful magnetic field since most of its fuel has decayed, and its atmosphere and prospects for future Martian life have since been stripped away.

Today, 99.3 percent of uranium is the fertile isotope U-238, and 0.7 percent is the fissile isotope U-235. This ratio results from U-238's half-life of 4.5 billion years, meaning since the Earth formed only about half has decayed. In contrast, U-235 has a half-life of 704 million years, meaning that every 704 million years half as much remains. The concentration of U-235 in the Earth is tiny since most has decayed. Thorium, with its half-life of fourteen billion years, is four times as abundant as U-238 in the Earth's crust. U-235 is "fissile," meaning that it likely fissions when it absorbs a neutron. In contrast, Th-232 and U-238 cannot sustain a chain reaction. However, these elements are "fertile," meaning that if they absorb a neutron they become fissile.

The implications for nuclear power, as a result, are that "breeding," or the conversion of fertile isotopes into fissile ones, is the key to all but inexhaustible energy. In light water reactors, only the U-235 is used as a fuel. As such, light water reactors use less than one percent of the potential energy of the uranium. However, in fast reactors, the U-238 is converted into plutonium and consumed. It is also possible to design a thermal neutron reactor that converts Th-232 into U-233, which is consumed. So basically, U-238 and Th-232 are fuels; they just need to be transformed into Pu-239 or U-233 first, respectively.

Light water reactors, the most common type of nuclear plant in use around the world today, uses neutrons that are moderated (slowed) by light atoms like hydrogen. Water under high temperature and pressure is used as both the neutron moderator, due to its hydrogen content, and as the liquid heat transfer material that removes heat from the reactor and transfers it to a steam generator. The uranium 235 is "enriched" to a higher concentration than the 0.7 percent found in nature, usually to about 3-5 percent. The fuel rods containing the U-235 then undergo sustained fission for about three years. A fissile U-235 atom absorbs a neutron, becomes unstable and fissions, releasing fission products, heat, and two or three neutrons. These neutrons then collide with other uranium atoms, keeping the process self-sustaining. Some of the neutrons may be absorbed by the 96 percent of the uranium that is U-238, causing it to become U-239, which beta-decays into neptunium 239, which after 1.7 days further beta-decays into Pu-239. When the fuel rods are "spent," about 95 percent of the material in the fuel rods is U-238, one percent Pu-239, one percent U-235, and three percent fission products.

In the United States and in Sweden, the entire fuel rod is simply stored away. Elsewhere, it is reprocessed for more energy. In France, the PUREX process—Plutonium and Uranium Recovery by EXtraction—is used to recycle the fissionable plutonium and uranium into MOX fuel, or mixed-oxide fuel. The fission products are vitrified in the form of a durable glass matrix, and stored away. The U-238 is saved elsewhere as a future fuel. All of France's vitrified fission products are stored in a single room in La Hague. While the plutonium stored away in spent fuel rods takes "100,000 years" to decay away to natural background levels of radiation, the fission products are no more radioactive than naturally occurring uranium in the ground in less than 500 years.

In fast reactors, the neutrons are not moderated but are instead allowed to remain "fast," hence the name fast neutron reactor or fast breeder reactor. In contrast, light water reactors are called "thermal" neutron reactors. When a Pu-239 or U-233 atom fissions three neutrons are often released, especially with Pu-239. This allows for one neutron to collide with another plutonium or uranium atom, keeping the chain reaction going, and the other two to collide with fertile atoms. As such, it is possible to design a reactor configured so as to produce more fissile material than it consumes, which allows for the complete consumption of U-238 and Th-232 resources through nuclear alchemy (44). Not only does this make it possible to obtain over 100 times the energy per ton of uranium, since 99.3 percent is U-238, but it also makes it economically possible to exploit low-grade ores like uranium in granite or seawater.

There are 104 commercial pressurized water reactors in operation in the United States, and about 439 worldwide. While no new nuclear plants have been ordered and built from the ground up since 1973, and the last plant completed was Watts Bar in 1996, the United States today obtains twice as much power from nuclear as it did in the 1970s since plants are run more efficiently. The light water reactor may have been a success story, but not the breeder reactor. In 1983, the United States attempted to build its first large-scale commercial breeder reactor, the Clinch River breeder reactor, to be built near the Clinch River, Tennessee under operation by the Tennessee Valley Authority. A joint effort by the U.S. Atomic Energy Commission, the DOE and the U.S. electric power industry, the CRBR was to be a demonstration design of a new series of breeder reactors that would phase out coal (which later, of course, ruined the Clinch River because it was not phased out).

Initial appropriations began in 1972. The reactor was to be cooled by liquid sodium metal and fueled by a mixed-oxide fuel form of uranium and plutonium. Sodium, unlike water, keeps neutrons "fast," allowing sufficient neutrons to pass through the coolant and collide with a breeder blanket of depleted uranium, or U-238 left over from the enrichment of U-235 for light water reactors. Thus, the reactor can produce more plutonium than it consumes, which can then be reprocessed for more energy and to breed yet more fissile material. Sodium metal was chosen because it not only keeps neutrons fast, but is also an excellent heat transfer medium due to a high specific heat, and is a liquid between 98°C and 892°C, an optimum temperature range for efficient electric generation. Sodium is a liquid at these temperatures at near atmospheric pressure; therefore no pressure vessel is required like in LWRs. Sodium is also extremely noncorrosive with steel. The downside to sodium, however, is that it reacts with water to release hydrogen, which is extremely reactive and explosive, so the sodium must always be contained behind steel, which is not a major problem. As a safety precaution, an intermediate sodium coolant loop between the reactor and the steam turbines is standard in breeder reactors, which increases costs.

As a result of the need for an intermediate coolant loop and the cost of reprocessing, the CRBR was more expensive than light water reactors at the time. It was determined that uranium, then $25 a pound, would have to reach $165 a pound until breeder reactors became commercially competitive. Another problem was proliferation, since breeders produce weapons-grade plutonium, which then needs to be manually reprocessed. In April 1977, President Carter called for a moratorium of construction of new breeder reactors, arguing that they would lead to proliferation if other nations followed suit, announcing "we will defer indefinitely the commercial reprocessing and recycling of the plutonium produced in the U.S. nuclear power programs." Two years later, Carter proclaimed reprocessing is "an assault on our attempts to control the spread of dangerous nuclear material," and vetoed funding for the CRBR, which ultimately culminated in Congress deciding to terminate funding in 1983, before the reactor could be completed.

Around the same time period, France attempted to build its own large-scale commercial fast breeder reactor, the Super Phoenix, near the Rhone River. With few natural resources of its own and worries of a shortage of uranium supplies for LWRs (huge reserves in Australia and Canada had not yet been discovered), France embarked upon the plutonium economy. Design work began in 1968, and construction lasted from 1974 to 1981, with power production up and running by 1985. The design was very similar to the CRBR, a loop-type sodium cooled plant fueled by MOX with a breeder blanket of depleted uranium.

On January 18, 1982, an environmentalist group attacked the power plant by launching rockets against it. Those responsible referred to themselves as an "eco-pacifist" group. Five rocket propelled grenades were launched at the incomplete containment building, two of which hit and caused damage, both narrowly missing the empty reactor core which could have been seriously damaged. So much for "eco-pacifism," when humanity's last best hope at avoiding economic and environmental ruin has missiles launched at it by the peaceful "greens."

The French green party, Les Verts, hated the Super Phoenix even more than LWRs. Eventually it helped form the national network Sortir du Nucleaire through the unification of hundreds of antinuclear organizations to oppose the power plant. In 1997, following a court case brought forth by opponents of the power plant, a French court, Conseil d' Etat, ruled that a 1994 decree authorizing the Super Phoenix to continue producing power was invalid. In June 1997, Prime Minister Lionel Jospin finally closed the power plant, arguing that it was expensive. However, Jospin's government included many "green" ministers opposed to nuclear power, and many have argued that the decision was political. Nevertheless, like the CRBR, the Super Phoenix was more expensive and would create a proliferation hazard if built in unstable nations.

In 1989, Alvin Weinberg, inventor of the LWR, would write a memoir titled "The Second Fifty Years of Nuclear Fission," commemorating the fiftieth anniversary of Hahn and Strassman's discovery. In it, Weinberg remarked,

> The two primary aims of nuclear power—inexhaustible energy, i.e. breeding, and economically competitive electricity—have both been demonstrated; what has not been demonstrated is electricity in a large-scale breeder that is cost-competitive today. The largest breeder, the 1200-MW Super-Phenix, is too expensive. Super-Phenix has placed a cap on the cost of electricity for many future generations, if not for millennia, and this cap is surely less than twice current costs of electricity. Whether fusion or solar electricity will match this, and will eventually displace the breeder, remains to be seen (17).

Nuclear power was in somewhat of a bind. The light water reactor had demonstrated very cheap electricity, as cheap as coal or hydropower in most places. However, for the LWR uranium supplies will last at the most a few centuries without unduly raising the cost of electricity as high-grade ores are depleted. Supply constraints may occur even earlier as the world rapidly builds more reactors to displace fossil fuels and allow future economic growth. LWRs also produce long-lived waste, since if plutonium is not recycled it takes 100,000 years to mostly decay. In contrast, breeders are unlimited by fuel supplies, but are more expensive and a proliferation hazard. What the world would need, if solar or fusion could not meet the challenge for the foreseeable future, was a perfect breeder.

Weinberg also remarked,

> As the one who was involved in the original decision to power NAUTILUS with a light water reactor, I have never outgrown my astonishment that LWR became the dominant reactor. After all, light water was chosen originally for submarines because such reactors are compact, and at least in principle, relatively simple ... (LWR) was chosen for Shippingport after President Eisenhower had vetoed the Navy's proposal to build a nuclear aircraft carrier powered by a larger version of the NAUTILUS power plant.

The first atomic submarine, the USS *Nautilus*, was powered by a compact LWR. Plans for powering aircraft carriers using nuclear fission were canceled until a later date, and as such the first commercial power plant in the United States, the Shippingport reactor, was opened in 1957, powered by the light water reactor that would otherwise have been used for aircraft carrier propulsion. Due to the early introduction of the LWR and resulting improvements that followed, and the relative abundance of cheap uranium, the LWR won out. But as demand for power around the world grows, and breeders are eventually needed, how will the world deal with proliferation? Safety is also a concern, given meltdowns like Three Mile Island. While there is no accepted evidence that this meltdown harmed the public, it was still a massive financial loss. Weinberg concludes,

> Can we develop nuclear reactors whose safety is deterministic, not probabilistic, and which, if developed, would meet the public's yearning for assurance of safety? ... What the public's reaction will be, surely depends on the alternatives to, and upon the incentives for, nuclear power in the next 50 years. As for incentives, greenhouse may be the key ... We nuclear engineers of the first nuclear era have had success, yes, with our 500 commercial reactors, and our practical breeders. But the job is only half finished. The generation that follows must resolve the profound technical and social questions that are convulsing nuclear energy.

The Nuclear Economy

Nuclear power may indeed be all that stands between what we identify as civilization and its alternatives. Solar and fusion are simply nowhere near where they need to be, particularly solar, although nuclear fusion *may* come on-line mid-century. But, as House representative Roscoe Bartlett puts it, "counting on fusion to solve your energy problems is like counting on winning the lottery to solve your financial problems." Weinberg was correct that the greenhouse effect would be the key to triggering greater interest in nuclear energy. Scientists knew all along that nuclear energy would likely be needed one day, since there was no current sustainable alternative adequate for civilization's needs. Sadly, however, politicians chose to listen to the "greens" rather than the scientists, and nuclear power would have to patiently lie in wait until a fossil fuel crisis arose. Scientists realized that this would give them time to perfect nuclear power plants.

Would scientists be able to develop a practical breeder that is proliferation-proof, produces dramatically less waste if any at all, is cheap, is walk-away safe, and retains the blessing of unlimited clean energy? Can this be done in time to prevent potential social, political, economic and environmental upheaval as fossil fuels run short and contribute to greenhouse? As it turns out, shortly after Weinberg's writings, scientists at Argonne National Laboratory in the United States succeeded in doing exactly that beyond their wildest dreams—only to have their dreams shattered.

Chapter 23

New and Clear Nuclear

"We will terminate unnecessary programs in advanced reactor development," President Clinton remarked in his 1994 State of the Union address (18). His words will forever live in infamy. Little did Clinton realize just how necessary advanced reactors would be, and how unwitting his remarks would later seem given the looming converging catastrophes of peak oil and climate change. According to Dr. Charles Till, who led a team of 500 Ph.D.s who worked on a revolutionary new project,

> In the decade from 1984 to 1994, scientists at Argonne National Laboratory developed an advanced technology that promised safe nuclear power unlimited by fuel supplies, with a waste product sharply reduced in both radioactive lifetime and amount. The program, called the IFR, was cancelled suddenly in 1994, before the technology could be perfected in every detail. Its story is not widely known, nor are its implications widely appreciated (45).

The IFR, or Integral Fast Reactor, was a revolutionary fast reactor design that promised to completely eliminate all past problems associated with nuclear power. It promised unlimited clean energy, cheap, and walk-away safe—all while contributing nothing to proliferation. According to Till,

> (The name) Integral was chosen to denote the fact that every element of a complete nuclear power system was being developed simultaneously, and each was an integral part of the whole: the reactor itself, the process for treatment of spent fuel as it is replaced

by new fuel, the fabrication of the new fuel, and the treatment of the waste to put it in final form suitable for disposal . . . nothing was to be left behind to be developed later. No detail was to be left hanging, unresolved, to raise problems later, as had been the case in the present generation of nuclear power (45).

The Integral Fast Reactor, which was at the time the biggest and most heavily funded energy research project in U.S. history, truly provides exactly what humanity needs: electrons that are completely abundant, reliable, cheap, and clean. None of the other alternatives meet all four categories. Fundamentally different than fast breeders of the past, the IFR is both a fast breeder and fast burner of nuclear fuel, hence simply the name "fast" reactor. It segregates and consumes plutonium at the same rate it is produced, and mixes materials with plutonium in such a way that there is no proliferation hazard. The IFR is also sometimes referred to as the ALMR, or Advanced Liquid Metal Reactor.

The reactor, which was virtually complete when its funding was canceled, had benefits that included the following:

(1) The IFR contributes nothing to proliferation, because its pyrometallurgical fuel cycle with electrorefining mixes materials in such a way that they are dangerous to handle and not suitable for bomb production.
(2) It can be fueled entirely by material recovered from today's used nuclear fuel.
(3) It consumes virtually all the long-lived actinides that worry people who are concerned about "the nuclear waste problem," reducing the needed waste isolation time to less than 500 years.
(4) It could provide all the energy needed for centuries feeding only on U-238 that has already been mined.
(5) It utilizes uranium resources with 160 times the efficiency of today's LWRs.
(6) It has an unlimited fuel supply using low-grade ores.
(7) It does not require enrichment of uranium.
(8) It provides 24/7 base-load power.
(9) It could be emplaced in excavations at existing coal plants and utilize the same turbines, condensers, and electric grid infrastructure coal plants currently use.
(10) It can be built to phase out coal quickly, since it requires no pressure vessel during construction, and can be built into the grid decommissioned coal plants once used.

(11) Its power is inexpensive, between two and five cents per kilowatt-hour.
(12) It is passively safe from meltdown based upon the physical properties of the reactor, which causes the reactor to shut itself down when it overheats.
(13) Modular design: the IFR will likely cost less to build than LWRs.
(14) It is emissions free and does not contribute to climate change (18).

Now what's wrong with that? Absolutely nothing, which is why its termination was so preposterous. Especially given the fact that the Japanese wanted to chip in $60 million to fund the final stages of IFR development, so it actually cost more to terminate the project and decommission the prototype IFR than it would have cost to allow the project to be finished! (46)

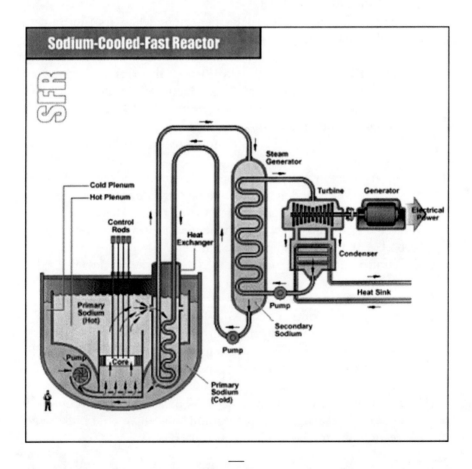

Why in the world was this reactor shut down? There are many possible explanations. Some feel that the fossil fuel industries may have had a role. Clinton's secretary of energy, Hazel O'Leary, may have been wary about the threat the IFR could have posed to the fossil fuel industries, particularly coal and natural gas, which are used for power generation. O'Leary was previously a lobbyist for fossil fuel industries, and served as executive vice president of the Northern States Power Company from 1989 to 1993, which used primarily natural gas. Natural gas prices, since the cancellation of the IFR, would increase by over a factor of ten by July 2008. As Steve Kirsch, respected Silicon Valley philanthropist and entrepreneur puts it,

> Why do you think the government would pour billions of dollars into the biggest energy research project in history and then not just cancel it, but do their best to bury it? The researchers at Argonne developed a safe and economical source of unlimited clean energy. Between that and the other renewable power technologies we wouldn't need oil, coal, gas or uranium mining/drilling anymore. We're talking about putting the most powerful corporations on the planet out of business. Not out of malice or spite, but simply because they won't be needed anymore and because what they're doing to the planet is killing us (18).

During senate testimony on whether or not to kill the IFR, upon questioning by Senator Paul Simon, Hazel O'Leary was forced to admit that because the Japanese were willing to chip in $60 million to finish the project, it actually would have been less expensive to finish the IFR than to terminate it. Not only was the IFR in the end terminated, but the scientists who worked on the project were also forced to remain silent about their work and not publicize it. During senate testimony, Senator Moseley Braun remarked,

> Secretary O'Leary this year awarded the general manager—and this is almost a funny story—the general manager of the ALMR/IFR program a gold medal and $10,000 for his work on this technology, and the secretary at the time described the ALMR/IFR as having "improved safety, more efficient use of fuel, and less radioactive waste." So why would the administration award someone $10,000 and a gold medal for a program that they then turn around and want to kill?

Clearly, something not right was at play. The fossil fuel industries may well have influenced O'Leary to convince Clinton the IFR was "unnecessary" and should be terminated, taking advantage of antinuclear sentiments on the left. Why else in the world would you terminate something that you have given awards to when it would actually cost more to do so?

If antinuclear sentiments were taken advantage of, doing so apparently wasn't very difficult. As Dr. Charles Till remarks about the final stages of IFR development,

> From the beginning, midway through the first Reagan Administration, success after success in the development work had allowed a broad and comprehensive program to be put in place. Every element and every detail needed for this revolutionary improvement in nuclear power was being worked on. (It was believed) another two years should bring successful completion . . . (By 1993) the new Clinton administration had brought back into power many of the best-known antinuclear advocates . . . Democrats were in the majority in both houses of congress. Antinuclear advocates were also settling into key positions in the department of energy, the department that controlled IFR funding. Other antinuclear people were now in place in the office of the president's science advisor, in policy positions everywhere in the administration, and in the White House itself (45).

Dr. Charles Till was desperate to finish the project, nearly complete, that had consumed a decade of work by 500 highly skilled and motivated professionals. Upon meeting with the deputy director of the White House's Office of Science and Technology Policy to plead his case, Till was told: "No, it has to go. It's got to be shut down. It's a symbol." Clearly, the unwitting ideology and religion of the antinuclear crowd had permeated deep within the Clinton administration and doomed the IFR. Not only was the project terminated, but the prototype IFR was decommissioned, which will mean a whole new one will need to be built to demonstrate the final details of IFR performance before commercial IFRs can be built.

While the egregious decision by the Clinton administration to cancel the biggest energy research effort in U.S. history solely out of political disdain seemed to doom the IFR at the time, there was a chance that if both the House and Senate voted to fund the IFR, the project could continue. In the Senate, which ultimately voted to support the IFR, democrat Bennett Johnston of Louisiana led the pro-IFR forces. As chairman of the energy and water subcommittee, which oversaw IFR funding, Johnston countered many of the anti-IFR arguments, which were led by Senator John Kerry.

Sadly, however, the House had no experts on energy or nuclear power to defend the project, and many were misinformed that the IFR would be an expense to finish and a "breeder" that causes weapons proliferation. As a result, the House voted to cancel the IFR, while the Senate supported it. The decision therefore went to the administration to evaluate behind closed doors in conference. The plug on the IFR, in the end, was pulled. The U.S. government would then proceed to pour billions of dollars into corn ethanol and hydrogen research, and the gradual decline of one of the great empires of the world would ensue. Would U.S. civilization one day collapse when oil runs short, as did Mayan civilization due to depletion of wood resources? Little did the politicians who cancelled the IFR realize the implications of their decision.

Some House members later realized that they had been duped and expressed regret that they had made the wrong decision. According to Congressman Myers of Indiana, "The thing that really bothers me is that after the vote, a number of members came up to me and said, 'We voted wrong. We thought we were saving money.'" But by the time they realized that they had been duped by the anti-nuclear forces, it was too late. The IFR was finished. Upon Dr. Till's retirement in November 1997, Senator Kempthorne would remark,

> I rise today to mark the retirement and celebrate the career of one of our nation's great leaders in science . . . his greatest contribution, to both his discipline and to the world, was in the development of the Integral Fast Reactor, the IFR. This inspired source of electrical power has the capacity to achieve incredible efficiency in fuel use, while significantly lessening problems associated with reactor safety . . . Unfortunately, this program was cancelled just 2 short years before the proof of concept. I assure my colleagues someday our nation will regret and reverse this shortsighted decision. But complete or not, the concept and the work done to prove it remain genius and a great contribution to the world (47).

President Bush would go on to restart the IFR program in 2003, renamed "The U.S. Generation IV Fast Reactor Strategy (48)." It would incorporate advanced sodium-cooled reactors like the IFR, as well as advanced lead and helium gas-cooled models that offer the possibility of even higher thermal efficiencies. Sadly, due to the IFR's cancellation, large-scale chemical and metallurgical electrorefining and pyroprocessing has not been demonstrated, and commercial IFRs could be built no earlier than about 2015 (14). By then, hopefully a large-scale demonstration plant will be up and running, and the world will begin rapidly building IFRs. Until then, improved and proven Gen-III+ light water reactors will need to be built, and their waste, which can be stored right next to the reactor for decades, will eventually be used to fuel future IFRs.

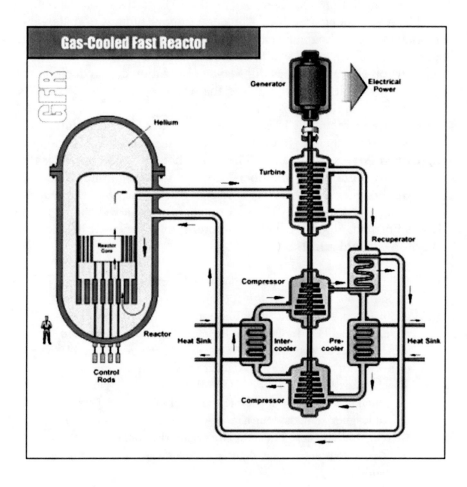

Chapter 24

Weapons Proliferation

When debating the fate of the IFR on the floor of the U.S. Senate, Senator John Kerry argued that the IFR project would be too expensive, would damage the environment, and would cause proliferation. In reality, the complete opposite of all three positions is true. According to Kerry,

> The advanced liquid metal reactor is an expensive pork-barrel project that poses serious environmental and proliferation risks... breeders convert uranium into plutonium, the material used to make nuclear weapons. By promoting a fuel cycle based on plutonium, the ALMR inevitably increases the risks of nuclear proliferation.

Kerry starts out by saying the IFR "is an expensive pork-barrel project," when realistically, it would have been cheaper to complete the IFR. Powering the country on coal as an alternative to the IFR would go a long way toward increasing global warming, so the "serious environmental" risk argument can also be refuted. It is ironic that Kerry, now a champion of fighting global warming, may be one of the U.S. senators unwittingly most responsible for exacerbating it by defeating our one good alternative to fossil fuels. But wait, the reactor "promotes a fuel cycle based on plutonium," so wouldn't there "inevitably" be a proliferation risk? A careful analysis reveals that not only does the IFR contribute nothing to weapons proliferation, it actually greatly reduces proliferation risks LWRs would pose.

LWRs are themselves breeders of plutonium. While there is a net consumption of fissile material, some plutonium is still bred inside the reactor. In France, this plutonium is extracted using the PUREX process, which does indeed isolate plutonium of sufficient weapons-grade quality, albeit poor quality due to the trace presence of Pu-240 and 241. In the United States we simply store away our spent fuel rods without recycle, but this could one day lead to future proliferation risks since the plutonium, mixed in with highly radioactive Sr-90 and Cs-137, is protected from extraction only temporarily by the presence of these fission products. Today the PUREX process uses advanced technology to extract plutonium, done behind shielding. However, after a few hundred years, fission products will mostly have decayed, making it easier to mine for plutonium buried in geologic repositories, since the spent fuel wouldn't be hot enough that you would need to be behind shielding to handle it. In that sense, even if you don't recycle, plutonium mines could be created by simply burying away our spent fuel.

IFRs, in contrast, can eliminate this risk by burning used LWR fuel *without separating plutonium.* Bomb makers demand plutonium of at least 94 percent Pu-239 isotopic quality (46). The IFR fuel cycle, which does not require PUREX extraction, uses only lower-grade plutonium with hot fission products and actinides of similar isotopic mass mixed in. According to George Stanford, one of the reactor physicists that worked on the IFR, the IFR fuel cycle was evaluated by expert bomb designers at Livermore National Laboratory:

> They looked at the problem in detail, and concluded that plutonium-bearing material taken from anywhere in the IFR cycle was so ornery, because of inherent heat, radioactivity and spontaneous neutrons, that making a bomb with it without chemical separation of plutonium would be essentially impossible—far, far harder than using today's reactor-grade plutonium (46).

But if there is plutonium, could it be possible to use chemical separation to obtain material suitable for a bomb? Stanford explains,

> First, you would need a PUREX-type plant, which doesn't exist in the IFR fuel cycle. Second, input material is so fiendishly radioactive that the processing facility would have to be more elaborate than any PUREX plant now in existence. The operations would have to be done entirely by remote control, behind heavy shielding, or the operators would die before getting the job done. The installation would cost millions, and would be very hard to conceal. Third, a routine safeguards regime would readily spot

any such modification to an IFR plant, or diversion of highly radioactive material beyond the plant. Fourth, of all the ways there are to get plutonium—of any isotopic quality—this is probably the all-time, hands-down hardest (46).

So, in other words, if you have the money and technology to actually build a PUREX plant more elaborate than the ones now in use, you would already have the resources to build atomic weapons anyway. In a sense, this is because IFR fuel is dirty while IFR waste is clean—the opposite of clean LWR fuel and dirty LWR spent fuel. The problem of plutonium and management of other actinides is turned into a solution; long-lived actinides are consumed during the reaction for 160 times more energy per ton of uranium, their similar weight impedes centrifugal separation, and the short-lived fission products that remain serve to aid proliferation resistance by keeping IFR fuel hot. Kerry's logic that there is an increased proliferation risk simply because plutonium is used as fuel in IFRs is simply false; proliferation risks are greatly reduced by getting rid of plutonium, and doing so with a hot fuel-form incompatible with nuclear weapons.

There are far, far easier ways to make a bomb than chemically purifying IFR fuel, which is not suitable for bomb production but works well for power generation. There is controversy that in Iran, uranium is being enriched to make uranium bombs. Iran, of course, simply claims it needs electricity. If naturally occurring uranium is enriched to over 90 percent U-235, it is possible to make a uranium bomb like the one dropped over Hiroshima. This is the easiest way to make an atomic bomb. And, since light water reactors require enrichment of uranium as part of their fuel cycle, the only way to avoid proliferation using LWRs is for a safeguards regime to supply a power-hungry rogue state with uranium already enriched to 3-5 percent U-235 for reactor fuel. Even then, the rogue state may simply invent relatively simple diffusion enrichment technology on its own. Since the IFR does not require enrichment of uranium, it could safely be built in rogue nations without proliferation concerns, and if a state then pursued the development of uranium enrichment technology, it would be obvious that this was being done solely to obtain a bomb.

Not only could IFRs be fueled by used nuclear fuel before it decays to the point that it becomes a proliferation concern, even down-blended fissile material from nuclear warheads can fuel IFRs. Today, half the nuclear fuel used in U.S. LWRs is down-blended warhead material. Since 20 percent of U.S. electricity currently comes from nuclear fission, one in ten light bulbs are therefore lit by warhead material. In other words, no matter where IFR fuel comes from, IFRs decrease the likelihood of weapons spreading around the world, rather than increasing such a risk.

In some impoverished nations, there is no existing electric grid. A third of the world's population still lives without access to electricity. In these nations, a proliferation resistant "nuclear battery" could be deployed, a type of self-contained nuclear reactor. After being manufactured at a central location and rented to nations needing more electricity, nuclear batteries could be buried underground, completely sealed, and provide reliable power for a period of fifteen to thirty years before needing to be refueled. At the end of their life span, the exhausted reactor cores, still sealed, could be traded for rejuvenated ones (49). Not only would trying to open a core violate the terms of agreements with power providers, but the fissile material in the core would also be low-grade and so hot that reprocessing would need to be done behind shielding with advanced technology, as with IFR fuel. Toshiba has offered to demonstrate its design of a ten MW nuclear battery in Galena, Alaska (population 650) for free. The design, termed the "4S," (Super Safe, Small and Simple) would be perfect for small, isolated communities living off the grid.

Al Gore, the most prominent champion of the fight against global warming, has apparently not been briefed on either IFR or nuclear battery technology, nor has he even bothered to familiarize himself with the latest in nuclear power plant technology (18). Given statements Gore made in a September 18, 2006 speech at NYU he apparently doesn't even remember his administration's termination of an advanced reactor designed to solve all of the problems associated with nuclear power, and he may be especially excited to learn about the nuclear battery:

> The main reason for my skepticism about nuclear power playing a much larger role in the world's energy future is not the problem of waste disposal or the danger of reactor operator error, or the vulnerability to terrorist attack. Let's assume for the moment that all three of these problems can be solved. That still leaves two serious issues that are more difficult constraints. The first is economics; the current generation of reactors is expensive, take a long time to build, and only come in one size—extra large.

Saying nuclear plants "only come in one size—extra large" is woefully uninformed. Nuclear batteries and nuclear-propelled ships may play an important role in the future filling in the gaps big reactors can't fill, in either remote locations or portable vessels. The reason why most nuclear plants are big, in contrast, is simply because this is the most economical way to efficiently produce mass electricity. It took France twenty years to make nuclear power its primary source of energy, so in that sense nuclear plants can be built and scaled in a reasonable period of time—the same can't be said for solar installations. France now has the cheapest electric rates in Europe, and is Europe's largest exporter of cheap electricity.

We're going to need not just more electricity in the future, but to simultaneously be phasing out existing coal plants as well, and the IFR happens to be perfectly suited to replacing the boiler of coal plants and being directly built into their existing grid infrastructure—making deployment cheap, fast, and efficient. Given that solar is still stuck at 0.3 percent of U.S. electricity and 0.1 percent of overall energy, making the claim that nuclear power plants "take a long time to build, and only come in one size" demonstrates a fundamental lack of knowledge about solutions to global warming. Gore then goes on to say,

> Secondly, if the world as a whole chose nuclear power as the option of choice to replace coal-fired generating plants, we would face a dramatic increase in the likelihood of nuclear weapons proliferation. During my eight years in the White House, every nuclear weapons proliferation issue we dealt with was connected to a nuclear reactor program. Today, the dangerous weapons programs in both Iran and North Korea are linked to their civilian reactor programs.

For something like the IFR to come along might seem too good to be true, but thankfully it solves all of the problems Gore brings up—most notably weapons proliferation. The biggest difference between earlier breeders and the IFR is its revolutionary new pyrometallurgical fuel cycle with electrorefining that was specifically designed to solve the issue of proliferation by not chemically separating plutonium. Apparently, Gore doesn't remember the IFR being killed when he was a part of the Clinton administration, as also demonstrated in a later interview between Gore and Idaho Senator Larry Craig (50). Nuclear batteries themselves are very proliferation resistant, and would be a perfect power source for developing nations where people still live off the grid.

Gore seems to think "every nuclear proliferation issue we dealt with" was somehow a technologically incorrigible problem for which his administration couldn't figure out a way to solve after eight years of struggling with the issue. Because his administration terminated the IFR, proliferation may now become a global crisis as energy-hungry developing nations without IFR technology are forced to pursue reactor programs on their own that still require uranium enrichment. If only IFRs were now being built around the world, future weapons proliferation problems could have been avoided. This book was written specifically to inform people, just like Gore, who have simply been led in the wrong direction, and are misinformed about the practical solutions to the energy problem.

Chapter 25

Waste

In 1972, French scientists exploring for uranium in West Africa noticed something odd about uranium recovered from a mine in Oklo, Gabon. Uranium is always 99.3 percent fertile U-238 and 0.7 percent fissile U-235, but this ore had less than half that amount of U-235. At first, scientists believed terrorists may have somehow enriched uranium and buried away the depleted uranium 238 from the enrichment process. Shortly, however, they realized they had discovered the remains of something truly spectacular—the waste left behind by natural light water reactors that operated within the Earth over a billion years ago (51).

French physicist Francis Perrin announced the discovery of the first and only natural nuclear reactors ever discovered. As it turns out, since U-235 has a 704 million year half-life, around 1.7 billion years ago the ratio of U-235 to U-238 was naturally 3-5 percent as opposed to the smaller 0.7 percent concentration found today. As such, uranium was naturally already at the right concentration to sustain the fission chain reaction found in today's light water reactors, which require uranium enrichment of U-235 to 3-5 percent for fission to occur. Water that naturally filtered down to uranium ores through sandstone rock acted as a natural neutron moderator, just as in light water reactors, slowing neutrons so that they could be absorbed by other U-235 atoms and allow a chain reaction to occur. When heat became too great, the water turned to steam, slowing the chain reaction, then, once the water cooled back to a liquid, the chain reaction could start up again. A total of fifteen natural light water reactors were found in three different ore deposits at the Oklo mine in Gabon; over their entire 1.7 billion year history their waste moved ten feet from where it was created (51).

Once the natural reactors exhausted their fuel, the wastes they generated were held in place deep under Oklo by granite, sandstone, and clays surrounding the reactors. Today, virtually all of the waste has since decayed to release heat, and only traces of elements like plutonium remain. Not only is nuclear waste a naturally occurring substance, nature can safely store and contain it, even for 1.7 billion years. The Oklo reactors have provided scientists with definite proof that it is possible to store nuclear waste safely over indefinite periods.

The argument is commonly made that it takes nuclear waste "100,000 years" to decay away. People imagine some green bubbling goo like on "The Simpsons" that leaks and gives fish third eyes. In reality, the 100,000 year figure refers to the metal Pu-239. The half-life of Pu-239 is 24,000 years, so most has decayed after 100,000 years. Far from "one of the most toxic substances known to man," as Ralph Nader has claimed, plutonium is an alpha emitter, and alpha waves can't penetrate a piece of paper or the thick outer layer of human skin that we have evolved to protect ourselves from natural background radiation. Pu-239 can be safely held in the human hand in order to feel its self-generated heat. This, however, is a proliferation risk—plutonium can be placed in one's pocket and someone could safely walk right out of a building with it. This is why manually reprocessing high-grade plutonium is a weapons proliferation danger, which the IFR eliminates.

Since the IFR segregates and consumes all plutonium bred from U-238 as fuel, only fission products are left behind. Some decay almost instantly. Other fission products, like Sr-90 and Cs-137, have half-lives of a few decades. They keep IFR fuel really hot, so nothing can come close to it safely without shielding. These fission products also have certain medical and industrial uses, including radiotherapy for cancers, food irradiation to kill bacteria, and energy to power small devices like Sr-90 pacemakers or sensors in Cs-137 smoke detectors. Today the United States imports most of the fission products it uses for medical and industrial applications from Canada since reprocessing is discouraged.

Because fission products are so hot, IFR waste (or LWR waste that has had its plutonium and uranium removed through PUREX) decays to the point that it is no more radioactive than background levels of radiation in a few hundred years. At this point, all that's left is weakly radioactive fission products and traces of hot ones. As explained by Tom Blees, author of "Prescription For The Planet," an excellent book about the IFR,

> Once uranium and/or plutonium (from spent fuel or decommissioned weapons) enter the IFR power plant, it never emerges except in

the form of electricity and a small amount of fission products. Within a few hundred years, the radiation from this waste will be below normal background radiation levels. Yet it will exit the plant entombed in borosilicate glass, which will resist leaching any of it into the environment for thousands of years, long after any radiation has dissipated (14).

Thanks to the almost unimaginable energy density of nuclear fuel, the amount of waste a one GW IFR produces would be just one ton per year, or about enough to fill a small filing cabinet. Even for unprocessed LWR spent fuel, it is just twenty-five tons per year—enough to fill the back of a pick-up truck, although it would flatten the truck, it is so dense. Per person, for an entire lifetime of energy from the IFR, this would be an amount equal to less than a half a ping-pong ball. That's not a lot of waste, and it is vitrified in the form of a durable borosilicate glass matrix where it can't harm anyone. Making the deal even better, IFR fuel can be recovered from today's used nuclear fuel from LWRs.

After the fission products have decayed, all that remains is traces of I-129 and Tc-99. As Blees explains,

> These are so-called soft beta emitters that can be incorporated into the vitrified waste stream and would pose no health hazards in that form, being bound up for thousands of years... (We could store them by) dropping the shielded canisters of vitrified waste into deep ocean areas with deep muddy bottoms. The canisters would bury themselves in the mud, and during the thousands of years that they would sit there without any chance of leaching into the environment even more mud would build up above, incorporating the vitrified substance and the casks containing it into the Earth itself (14).

No Yucca Mountain repository needed. If something is two million times as energy dense it produces two million times less waste, and as the Oklo mine demonstrated in Africa, it is possible to store this shockingly tiny amount safely. Blees offers an interesting ocean repository idea, but land-based repositories like Oklo are also possible. Either way, in as little as ten years scientists may have the technology to use fast neutrons given off by plutonium to transmute I-129 and Tc-99 into truly innocuous compounds. This has already been demonstrated in Hanford's fast-flux facility, and may be incorporated into future Gen-IV advanced fast reactors cooled by helium gas. As such, nuclear fission will likely become totally waste free.

Sometimes the argument is made that reactors contribute to global warming by giving off heat, which has been scientifically discredited. Global warming is attributed to greenhouse gases like carbon dioxide or methane, which reflect back heat as it radiates back out into outer space. Heat given off by nuclear reactors, or any thermal plant for that matter including fossil fuels, simply radiates back out into outer space. Just as how land cools at night after the sun has warmed the Earth, so too does heat from thermal plants dissipate, then ultimately radiate out into outer space. Compared to the amount of energy that reaches the Earth from the sun, heat given off by thermal plants is thousands of times smaller. As a result, virtually all of the energy trapped by the greenhouse effect will be from the sun, so the only way to decrease global warming is to get greenhouse gas levels under control.

The argument is also sometimes made that the nuclear fuel cycle contributes significantly to greenhouse gas emissions. It is true that mining and enrichment of uranium requires energy, often from fossil fuels, as well as construction of nuclear plants. However, the overall carbon footprint of nuclear is just six grams of CO_2 per kilowatt-hour, compared to over 900 grams of CO_2 per kilowatt-hour for coal (52). For natural gas it is 465 grams of CO_2 per kilowatt-hour, and even for photovoltaic solar sixty grams of CO_2 per kilowatt-hour, or over 100 grams of CO_2 by some estimates (52). Today, uranium mining is performed using insitu leaching, where a mine is simply flooded, then "yellowcake" uranium is absorbed from the water. The process has a very small carbon and environmental footprint, especially because one ton of uranium contains as much energy as three and a half million tons of coal, so the amount of material mined is exceedingly small. If uranium enrichment is performed using centrifuges, this requires twenty to fifty times less energy than diffusion uranium enrichment. Of course, IFRs would eliminate the need for mining, milling and enrichment of uranium.

Chapter 26

Cost

The biggest cost for nuclear is simply building the power plant. Even for light water reactors, by far, most of the cost is getting a plant built, licensed, and operating. Uranium is so energy dense, fuel costs are very tiny. In contrast, for natural gas plants most of the cost is buying gas, as it is for coal to a reduced extent. IFRs, of course, would use fuel that is free—in fact, it would be better than free, since we could recover it from our nuclear waste stockpiles which we "don't know what to do with." As a result of very low fuel costs, once they are built nuclear power plants produce very cheap electricity. According to the Nuclear Energy Institute, in 2005 the average cost of producing nuclear electricity in the United States was 1.72 cents per kilowatt-hour, compared to 2.21 cents per kilowatt-hour for coal, 7.51 cents for natural gas and 8.09 cents for oil. By 2008 this had jumped to 1.87 cents per kilowatt-hour for nuclear, 2.75 for coal, 8.09 for gas and 17.26 for oil. So, once a nuclear plant is built, it is actually cheaper than fossil fuels. This difference is likely to increase even more dramatically in the future if there is a carbon tax, and as fossil fuel prices—already very volatile for natural gas and oil—continue to rise. Hawaii still generates most of its electricity from oil, and could greatly benefit from nuclear battery plants.

Today, construction of a new Gen-III+ light water reactor costs about $4,000 per kilowatt of capacity. This means that the cost of a one GW nuclear plant would be about $4 billion, or roughly twice as expensive as a dirty pulverized coal plant that spews millions of tons of CO_2 into the air a year. Nuclear is emerging as the cheapest overall energy source (other than hydro, which is maxed out). Nuclear plants produce very low-cost, low-pollution, reliable energy for over sixty years once built, so they pay for themselves overtime. Coal plants, in contrast, last only thirty years, and windmills fifteen to twenty years. As fossil fuel costs rise and the likelihood of a carbon tax is taken into account, nuclear emerges as by far the cheapest expandable source of electricity.

Nuclear was the cheapest overall source of power in the 1960s, but regulatory changes and delays quadrupled the cost of building nuclear plants by the early 1980s, leading to the misconception that nuclear is expensive. Today, improved and simplified designs, as well as changes to the licensing process of operating a nuclear plant, have helped make nuclear competitive. However, no nuclear plants were built in the United States start to finish that were ordered after 1973, and it wouldn't be until 2007 that NRG energy would apply to begin building new reactors. What went wrong in the late twentieth century?

In the early 1970s, a typical reactor cost $170 million, but by 1983 the cost had reached $1.7 billion, a ten-fold increase (53). Even taking inflation into account, this was still a four-fold increase. As illustrated by nuclear physicist Bernard Cohen,

> Commonwealth Edison, the utility serving the Chicago area, completed its Dresden nuclear plants in 1970-71 for $146/kw, its Quad cities plants in 1973 for $164/kw, and its Zion plants in 1973-74 for $280/kw. But its La Salle nuclear plants completed in 1982-84 cost $1160/kw, and its Byron and Braidwood plants completed in 1985-87 cost $1880/kw—a 13-fold increase over the 17-year period (53).

This was partly due to rising labor, steel and concrete costs, as well as inflation. But it was mostly due to stiff new regulations imposed on nuclear plants by the Nuclear Regulatory Commission and its predecessor, the Atomic Energy Commission Office of Regulation, beginning in the early 1970s. These stiff new regulations, known as "ratcheting," increased the amount of concrete used in new nuclear plants by 27 percent, the length of electrical cable 36 percent, steel 41 percent, and lineal footage of piping 50 percent by the late 1970s (53). The measures, which were mostly the result of safety concerns by an uninformed public, not only did nothing to improve plant safety—they may have actually decreased it. Extra piping, if not necessary, just means all the more piping that may fail as in Three Mile Island. Today's Gen-III+ and Gen-IV plants have simplified designs with less piping and materials costs like original plants, while also incorporating innovative new passive safety features, making them super-safe while still cost competitive.

Not only did "ratcheting" increase materials costs, but it extended the amount of time required to build a nuclear plant, leading to further cost increases. The time it took to construct a plant rose from seven years in 1971 to twelve years in 1980 (53). Even worse, plants were effectively required to get two licenses, one for construction and one for operation. After construction, which itself was subject to numerous safety analysis and NRC hearings, a separate permit to begin operating the plant was required. Public hearings would then begin anew as conflicts with groups opposed to the plant would be debated, and often as a result of those hearings, further changes would have to be implemented to a plant, increasing costs even more before a license could even be obtained to begin operation. This would later cause banks to not want to loan money for nuclear plant construction, and if a second license to operate was not given, billions spent on plant construction would be lost. It was almost as if the United States was trying to kill off its savior.

Shoreham plant on Long Island, NY, is infamous for producing the most expensive electricity ever generated. Built over the course of 1973 to 1984, the plant wasn't able to begin power production until April of 1989, due to lack of an "evacuation plan." Groups opposed to the plant were able to keep it from producing power even after it was already completed. They simply reasoned that because Long Island is surrounded by water, it would be more difficult to evacuate quickly if a plant meltdown occurred. Of course, a meltdown wouldn't even require evacuation due to the plant's containment shell isolating such an event. Governor Mario Cuomo refused to sign the proposed evacuation plan for the plant in June 1989, and the plant was forced to shut down after two months of only a small amount of noncommercial power generation, since it couldn't maintain a license to operate. Long Island now imports its electricity from coal and diesel powered equipment on the mainland, and has among the highest electricity prices in the Northeast. In 2005, two 100-foot tall wind turbines were built into the decommissioned plant's electric grid. They sporadically produce 1/8000th the amount of power Shoreham would have produced.

In Southern California, the two Diablo Canyon power plants were designed to withstand an earthquake of 7.5 on the Richter scale from nearby faults, such as the San Andreas. Construction began in 1968, but by 1973, another fault was discovered off the coast underneath the ocean floor. Before the plants could get licenses to operate, anti-nuclear protestors raised concerns that two simultaneous quakes might occur at once. In 1981, 1900 activists were arrested at Diablo Canyon, the largest arrests of any nuclear plant protest. Pacific Gas & Electric Company spent six years fighting hearings, referenda and litigation to have the plants approved, and was only able to do so after further modifications to the structure of the plants to make them resistant to multiple quakes.

By the 1990s, it was politically almost impossible for utilities to build more nuclear plants, so utilities were forced to focus on fossil fuels, primarily natural gas. However, natural gas prices are now extremely volatile, and global warming concerns have made coal politically more controversial than nuclear. In 2005, an energy bill designed to allow the construction of more nuclear plants in the United States solved the unpredictable licensing process by mandating a single license for construction and operation. Furthermore, Gen-III+ plants and IFRs will have uniform designs, like the EPR, AP-1000 or GE-Hitachi's economic boiling water reactor, which has been assessed to experience a meltdown once every twenty-nine million reactor years (54). Even if thousands of these plants were built, a meltdown would likely occur once every 10,000 years.

Of the 104 nuclear plants operating in the United States, there are eighty different designs. Having only a few super safe, streamlined, improved designs will make mass production much easier and faster, as will a predictable licensing process. Secretary of Energy Steven Chu has made expanding nuclear power an agency priority, and in June 2009 extended $18.5 billion in loan guarantees to four power utilities, Unistar, NRG, Scana Corp. and Southern Company (55). Loan guarantees will guarantee banks that if something goes wrong, the government will bail a utility out. This has allowed utilities to actually be able to once again get banks to finance constructing new nuclear plants, and will only be needed for the first few nuclear plants built to prove new designs are economical.

In April 2009, Senator Lamar Alexander called for the construction of 100 new nuclear plants in the United States. Worldwide, 45 Gen-III+ light water reactors are now under construction, including 388 planned or proposed (54). Nuclear power is truly experiencing a "renaissance." While most of the new plants built will be improved light water reactors over the next five to ten years, IFRs will likely follow, and will be able to eat the waste produced by LWRs.

It takes years of commercial scale operational performance to get a new nuclear plant design approved by the NRC, therefore to get commercial IFRs up and running will require building a demonstration IFR to prove large-scale commercial electrorefining technology. This technology already worked fine on a smaller scale, and certification of IFRs may be possible before 2015 (14). Getting the IFR certified must become an agency priority if the world hopes to rapidly expand nuclear power while solving all the problems associated with it.

The compact IFR may be cheaper to build than LWRs, since it requires no pressure vessel and uses a pool-type sodium coolant rather than the more expensive loop-type coolant of past fast breeders. Even better, the IFR can be deployed quickly into the infrastructure of existing coal plants, further cutting costs. As General Electric's Eric Loewen explained in a June 2009 conference about the IFR,

> They could be emplaced in excavations at existing coal plants and utilize the same turbines, condensers, and grid infrastructure as coal plants currently use ... Essentially all you'd be replacing is the burner. Thus you avoid most of the stranded costs. If stranded costs can thus be kept to a minimum, both here and, more importantly, in China, we'll be able to talk realistically not just about stopping to build new coal plants but replacing the existing ones, even the newest ones (56).

Loewen goes on to give the conservative estimate that IFRs would produce electricity at just five cents per kilowatt-hour, which would be competitive with IGCC coal even without a carbon tax. Costs would likely come down further in time, and using the existing grid infrastructure to rapidly phase out coal helps eliminate stranded costs. The path to saving the environment has been shown, now it must quickly be implemented. The Crystal River complex in Citrus County, Florida has four coal plants and one nuclear plant. If the four coal plants were converted to IFRs, the existing spent fuel stored on-site from decades of operation of the single nuclear plant could power the four IFRs into the twenty-second century.

Chapter 27

Radiation Controversy and Nuclear Plant Safety

People fear what they don't understand, and especially fear what they don't see and don't understand. Radiation can't be detected by any of the five senses even though it is all around us, and many of the fears related to nuclear power involve uninformed perceptions regarding radiation exposure. Massive doses of radiation are very destructive, but very slight doses shouldn't be feared. The public tends not to distinguish between high doses of exposure and very slight doses, to which life on Earth has adapted. If someone were to say that living near a nuclear plant exposes the public to radiation, this would immediately be viewed by most as harmful. However, when put in the context of exposing the public to less radiation than background radiation, nuclear plants seem less scary. When put in the context that living near a nuclear plant for a year exposes a person to less radiation than eating a banana, fear starts to dissipate. Understanding what radiation is, and what doses are acceptable is important for feeling more comfortable about nuclear power plants.

News and entertainment media have popularized the notion that radiation of any form or quantity is harmful, which seems intuitive. If massive doses can be lethal, an uninformed person might at first assume that any dose is harmful to some degree. This is known as the zero-threshold perception, which states that there are no effects at low doses that cannot be predicted from observations at high doses. However, scientists are becoming increasingly convinced that this theory is completely false, and that very low-dose exposure to radiation can even be beneficial to human health (57). Realizing that radiation isn't always some spooky ghost can be an important concept when it comes to understanding and accepting nuclear power, a concept that scientists themselves are already well familiarized with.

Known as radiation hormesis, the theory that low doses of radiation are beneficial could be analogized to stepping off a building versus stepping down stairs. Stepping off a building could be fatal, but simply stepping down stairs in incremental steps could provide beneficial exercise. Life on Earth evolved in a sea of radiation even higher than we are exposed to today, due to a higher concentration of radioisotopes like U-235 within the Earth in the past, which have since decayed. As a result, life had to adapt to survive in this sea of radiation, while only exponentially higher doses that life hasn't adapted to cause damage. This beneficial effect may be the result of very low doses of ionizing radiation inducing the production of special proteins that are involved in DNA repair processes, and the stimulation of the immune system (58). In 1909, it was demonstrated that mice treated with low-level radiation were more resistant against bacterial infection (58). In humans, scientists are increasingly finding radioactive foods to be healthful, like alcoholic beverages, antibiotics, bananas, and caffeine.

The human body itself contains about 8000 becquerels, or 8000 atoms disintegrating a second, due to the radioactive isotopes K-40 and C-14 that are found in food. Eating a banana exposes a person to 0.01 millirem of radiation due to the presence of K-40, while living near a nuclear plant for a year exposes a person to 0.009 millirem—even less. According to a Canadian study, the cancer mortality rate of actual nuclear plant workers is 58 percent lower than the national average (58). Similar studies in the U.K. have come to the same conclusion (58). The media has not reported this information widely, but probably would if nuclear plant workers had a greater rate of cancer (58).

What is radiation? Alpha radiation is a helium atom stripped of its electrons—two protons and two neutrons traveling at about one-twentieth the speed of light. Alpha radiation can't even penetrate the thick outer layer of human skin, but if an alpha emitter like plutonium is ingested and remains in the body a long time, extended doses can cause cancer to internal organs. Beta radiation is an electron traveling at 94 percent the speed of light. Beta particles can be stopped by tin foil, but can penetrate skin and cause reddening. Gamma radiation is pure energy traveling at the speed of light. It is very powerful, and is stopped only by four inches of lead or two feet of concrete. Cosmic rays are naked protons that rain down from outer space at half to 99 percent light speed, and may be the fallout from exploding stars.

The average American is exposed to 360 millirems of radiation a year. About thirty-five millirems come from cosmic rays, and forty-five from the geothermal processes of the Earth, mostly radon gas, a daughter product of uranium decay. The rest is mostly K-40 and C-14 within our bodies. Throughout the world, however, exposure levels vary widely. Granite has a high concentration of uranium and thorium, so living in New England or the Rocky Mountains tends to expose the public to more radiation than living in the Midwest or South.

Amazingly, people living in the Rocky Mountains, who are exposed to the highest levels of radiation in the United States, actually have 30 percent lower cancer rates than the national average. People living in the Louisiana Delta, with the lowest exposure rates, have the highest cancer rates in the nation. In Ramsar, a very radioactive region of Iran with many geothermal hot springs that emit radon gas due to uranium's decay, people are annually exposed to 18 rems—400 times the world average radiation exposure (59). People there seem to live perfectly healthy lives. Just as we are familiar with needing to "get some sun" and consider it healthy to have a slight tan, so too does the body need very minor doses of ionizing radiation to remain healthy, as this stimulates the immune system and DNA repair. The only difference is that we can see visible spectrum solar radiation, so we trust it; we don't realize that other forms of radiation are naturally all around us and also play a role. In that context, the 0.009 millirem a person is exposed to by living near a nuclear plant for a year is nothing to worry about compared to the 360 millirem the average American is exposed to each year due to completely natural factors.

Nevertheless, despite scientific data suggesting that very small doses of radiation are nothing to worry about, "greens" like Helen Caldicott and other antinuclear protestors continue to take advantage of an uninformed public. In a 2005 interview, Caldicott claimed cancer clusters are frequent among people living near nuclear power plants, "but because of the latent period of carcinogenesis, the incubation time for cancer is five to six years. You have to wait for a while and do a decent epidemiological study to access what's going on." To the average American Caldicott may appear scientific by the way she talks, but scientific tests have been performed the world over that just don't find this nonsense to ever be truthful.

In 1990, the U.S. Department of Health and Human Services, the Public Health Service, and the National Institute of Health undertook an extensive survey of cancer deaths between 1950 and 1984, in 107 countries with nuclear plants. Their three-volume study concluded that there was "no evidence to suggest that cancer mortality in countries with nuclear facilities was higher than, or was increasing in time faster than, the mortality experience of similar countries in the United States." In 1991, the National Cancer Institute determined after four years of research that there is no increased risk of cancer from living near nuclear facilities. Indeed, coal plants expose the public to 100 times more radiation than nuclear plants due to the presence of uranium and thorium in coal—very ironically—not to mention catastrophic climate change, acid rain, mercury poisoning and air pollution causing 24,000 deaths a year due to respiratory illness (42). Why it had to take catastrophic climate change in recent years to get the "greens" to start treating coal as harshly as nuclear boggles the mind. People need to learn to stop fearing anything "nuclear" the way our ancestors learned through science to stop fearing witches and warlocks.

The media has long championed the zero-threshold perception and the idea that U.S. nuclear plants can emit very high doses of radiation, creating a culture of ignorance and fear. In March 1979, the movie "The China Syndrome," starring Jane Fonda, got its name from the theory that a nuclear meltdown could melt through the containment structure of a nuclear plant "all the way to China." However, according to scientists in the movie, before this happened, the meltdown would hit the water table, vaporize, and render "an area the size of Pennsylvania uninhabitable." This was an eerily prescient comment, for two weeks after the movie opened, on March 28, 1979, at 4 a.m., a relief valve on a piece of equipment called "the pressurizer" would become stuck open at the Three Mile Island nuclear power plant *in Pennsylvania*. The power plant had just begun operation on December 30, 1978.

After a loss of coolant to the reactor, plant operators received conflicting data and incorrectly guessed that the plant was receiving too much water when in fact it was receiving too little. Plant operators, as a result, mistakenly turned off the reactor's automatic cooling safeguard, which caused a partial meltdown of a little over one-third of the reactor core. Hot metal hit cool metal, a heat transfer occurred, and the metal re-solidified. While far from "going to China," the incident was a huge financial loss.

Luckily, though, another fail-safe exists in light water reactors that can't be shut down—*a negative temperature coefficient*. Since the water served as both a heat removal medium and a neutron moderator, with a loss of coolant and overheating of water, the neutron chain reaction slows, shutting off further fission. If water coolant is lost and/or water turns to steam, the chain reaction stops. As a result, light water reactors cannot atomically explode based upon the physical properties of the reactor and the laws of physics. Another fail-safe is that in light water reactors, fuel rods must be arranged in precise geometric alignment so that the flying neutrons have a maximum opportunity of striking other uranium atoms. If the core starts to melt, fission stops, since this precise alignment is interrupted. As a result, there was only a partial meltdown, and not enough heat for the hot metal to melt through the containment dome.

As a fail-safe to avoid pressure buildups inside the reactor, operators vented a small emission of radioactive gas containing I-131. Over 2,000 personal injury claims were filed against the utilities running the plant, with plaintiffs claiming a variety of health injuries caused by radiation exposure. Over the next fifteen years, the case went to the Supreme Court and back. Finally, in 1996, district court judge Sylvia Rambo threw the case out, dismissing the lawsuit and granting summary judgment in favor of the defendants (60). While it was determined that plaintiffs were exposed to trace levels of radiation from TMI, there was no accepted evidence that this radiation dose was enough to harm the public, or any living thing in the vicinity (60). However, to the uninformed person, any radiation dose—no matter how miniscule—is often perceived as harmful, even if it is much lower than natural background radiation.

While a major financial loss, TMI was also a partial success story. No one died or was harmed, and the defense-in-depth design of the containment dome contained the meltdown. Actions such as better training of reactor staff were immediately implemented, and the accident galvanized scientists to design super-safe plants that rely on passive safety features not only for preventing a total meltdown by shutting down fission, but even for preventing such an accident from occurring in the first place. Sadly, the incident hurt the public's perception of nuclear power, and it wouldn't be until September 2007 that NRG energy would apply to the NRC to construct a new nuclear plant.

The situation at the Ukraine's Chernobyl plant, in contrast, would be another matter entirely. The monstrosity of a plant didn't even have a robust containment dome like western designs, and didn't have a negative temperature coefficient either. Whereas light water reactors naturally shut down fission due to loss of neutron-moderating coolant, the Chernobyl RBMK plant's neutron moderator was graphite. If graphite heats up, based upon its physical properties, it *accelerates* the speed of neutron fission, a dangerous positive feedback loop. If the graphite overheats, such a reactor can atomically explode. Fortunately, the rate of increase in fission isn't fast enough to reach atom bomb levels, but by the time it reaches TNT-levels, the whole plant can just blow itself apart, ending fission split seconds before an even more powerful explosion might have happened. This is exactly what occurred at Chernobyl.

In a RBMK style plant, the reactor core is a large block of graphite with holes in it containing tubes with fuel rods inside of them, and water flows rapidly through the tubes to remove heat. Since the neutron moderator is graphite rather than water, the water slows down excess neutrons and acts as a neutron "poison," actually preventing an increase in fission. Therefore, if there is a loss of coolant, the reaction can speed up, initiating more heat that burns graphite to cause more fission. This is a dangerous positive temperature coefficient, which no western nuclear reactor has. The only reason why RBMK plants were designed in such a way is that it requires less enrichment of uranium, and more U-238 in the fuel means more plutonium production for bombs. The soviets were using RBMK plants for simultaneous power generation and building up of their weapons arsenal.

On April 26, 1986, a team of operators at Chernobyl was working on supplying power to the grid as usual. Another was simultaneously running an experiment to determine whether momentum remaining in the turbines would be enough to power the cooling system during an accidental shutdown. An accidental coolant drop during the drill caused the reactor to overheat, sending a burst of steam instantly through the reactor turbines. A positive feedback soon melted the reactor core, causing it to mix with coolant, creating a burst of steam powerful enough to blow open the reactor. Even worse, the hot graphite burned for days, helping to disperse radioactive particulates into the air.

If such a reactor had a western style containment dome, that alone would have prevented emissions. Containment domes are extremely robust structures of concrete and steel. In the 1990s, the Air Force attached a Phantom F4 jet to a railroad track and rammed it into a concrete barrier, the thickness of a containment structure, at 500 mph. The jet was totally destroyed while the containment dome was only slightly dented. As a result, western nuclear plants are quite safe from terrorist attacks. Only an atomic bomb could pose serious harm to a nuclear plant, but even if terrorists obtained such a bomb, a densely populated city would be a much more likely target.

The fire in the plant was only extinguished by bombing the reactor with 5,000 tons of lead, sand, clay, and boron. Workers, many unprotected from high radiation levels, then built a steel and concrete structure over the reactor, which should have been there in the first place. During the three-month cleanup period of the reactor, forty-three workers died of radiation poisoning. About 4,000 Ukrainian children developed nonlethal thyroid cancer as a result of I-131 exposure, which migrates to the thyroid, while ten died unnecessarily from thyroid cancer that could have been treated. 16,000 people who lived within two to six miles of the reactor were exposed to fifty rem, and will likely have about a 4 percent higher chance of developing cancer (61). The city of Pripyat near the plant was exposed to 3.3

rems of radiation—still far less than the Ramsar region is naturally exposed to. As bad as Chernobyl was, the region is today far from uninhabitable, and is indeed a wildlife preserve since humans have all been scared away. The concentration of wildlife is many times greater in the region without the presence of humans, and as a result wildlife there is thriving (12). Ironically, as backwards as RBMK plants are, they are less of a threat to the environment than coal—the public just isn't educated enough to realize that scientifically. Coal plants spew out millions of tons of emissions a year as part of their normal operation, not accidents from obsolete designs.

In 1986, shortly before Chernobyl, engineers working at Argonne National Laboratory were testing revolutionary new "passive" safety features. Based upon the inherent physical properties of materials used and the laws of physics, such a reactor will naturally shut itself down without operator intervention or separate safety devices. As Dr. Charles Till describes,

> Amazingly, about a month before the Chernobyl accident, Argonne scientists had performed two remarkable demonstrations on their IFR test reactor in Idaho. An invited international audience watched the IFR shut down under accident conditions without any damage whatsoever. The first demonstration was precisely that of a Chernobyl-type accident as it might occur in an IFR. Then, on the same day, the exact conditions of the Three Mile Island accident were duplicated, again with a quiet, damage-free shutdown (45).

The first test performed on the IFR was a loss of flow test, simulating a complete blackout and loss of power to all systems. The second test simulated a loss of heat sink, in which the ability to remove heat from the plant by shutting off the secondary cooling system is disrupted. Normal safety systems were not allowed and the operators did not interfere. The EBR-II reactor on which the test demonstration was performed shut itself down flawlessly on both accounts. The name of this safety design, the S-PRISM, is engineered to be as flawless as possible technologically. It is so safe, even with thousands of IFR reactors built the odds of a Three Mile Island level meltdown occurring have been assessed at once every 435,000 years (14)!

The inherent safety of the IFR, based upon its physical properties, lies with the development of a revolutionary new fuel form, the metal-alloy fuel. This type of fuel responds to any event that could lead to significant accident by lowering reactor power levels, or even shutting the reactor down if necessary. Fast breeders like the Super Phoenix or MOX plants use a mixed-oxide fuel form, which is a poor heat conductor. Metal is a good heat conductor, which means that the interiors of the metal rods stay much cooler, and therefore far less heat is stored within IFR fuel. As a result, if there is a loss of coolant, far less heat is present, meaning the consequences of a hypothetical accident are less likely. In addition, as the fuel form becomes hotter than it should be, it starts to expand, disrupting neutron capture and slowing fission.

The reactor is also cooled by liquid sodium metal, a superb heat transfer medium that operates at near atmospheric pressure. As a result, no pressure vessel is needed, which makes exploding piping impossible. The IFR incorporates an innovative pool-type sodium arrangement, with the pool area itself covered in a blanket of nonreactive argon gas. Sodium's downside is that it reacts with moisture in air, and argon acts to prevent possible fires started by a possible sodium leak. Such leaks are unlikely, since sodium is extremely noncorrosive with steel. After thirty years of sodium use in the EBR-II experimental reactor, the welders' original markings were still visible on the joints that had been welded inside the reactor's piping.

Sodium is frequently used in industrial processes due to its superb heat transfer characteristics, and its challenges are well understood. There are 300 reactor-years of operational experience working with sodium in about eighteen fast breeder reactors that have been built worldwide, and the technology is well understood. In fact, the Russians have operated a commercial sodium-cooled fast breeder reactor in Beloyarsk, Russia, since 1980, the BN-600. It has the best operational safety record of supplying commercial power to the grid of any reactor in Russia's fleet, and a more sophisticated BN-800 model is now under construction. Sodium-cooled fast reactors and electrorefining are mature technologies, and the United States would be building IFRs today if not for the unfortunate turn of events that unfolded in 1994.

Conclusion

The shell must break before the bird can fly.

—Tennyson, The Charge of the Heavy Brigade,
Epilogue

In the face of economic and environmental collapse, and so-called "renewables" not measuring up, people are finally starting to realize the little secret that only nuclear power can revitalize the economy and environment. In 2008, at a dinner in Houston, former Chairman of the Federal Reserve Alan Greenspan said that "nuclear is the most reasonable power source," and "they have to use nuclear when all the trade-offs are made," drawing applause (62). The capitalist fundamentals of the U.S. economy remain unchanged, but peak oil has altered how they can function, and a massive new energy base must now rise to meet this unprecedented challenge while also being very low-carbon to avoid environmental collapse.

Nevertheless, the media still rarely discusses nuclear power, and television commercials sponsored by the oil and gas companies frequently boast ethanol, hydrogen, solar, wind, efficiency, etc. as being the answer. People in these commercials seem confident that current "renewable" technology will save the day, no matter how long it actually takes to implement. Likewise, politicians from both parties have failed the American people miserably on energy; from going to war in Iraq to canceling the IFR, it would seem that politicians want us to be stuck on fossil fuels as long as possible.

In June 2009, at an annual shareholder meeting in Dallas, ExxonMobil CEO Rex Tillerson said that the age of fossil fuels would last 100 years because there is no alternative (63). Conveniently, this is just long enough for most of the fossil fuels to be burnt, even coal. This book was written precisely to show that there is one alternative, which the fossil fuel industries rarely seem to mention. Petroleum man may be nearing extinction, but uranium man must rise to take his place. If we hope to preserve a habitable environment, we cannot simply burn all the coal only to turn to nuclear anyway after that. The fossil fuels companies are the largest companies in the world, doing trillions of dollars in business annually. They may lose a lot of money if their time comes a little sooner, but this wouldn't be due to malice or spite, just the fact that their services are no longer needed and are killing us.

Energy can be a tough subject to understand, and the fossil fuel industries have benefited from an uninformed public. A July 2009 poll revealed only 7 percent of respondents to be aware that electric power plants are the biggest cause of global warming, but when prompted 54 percent believed nuclear power plants are a major cause of global warming (64). The politicians who shaped energy policy unwisely chose to listen to the "greens" rather than scientists like Alvin Weinberg and Charles Till, and the public was too uninformed to object to energy policy based upon sources that at passing glance appeared promising. Now that it has been demonstrated through decades of trial and error that so-called "renewables" can't possibly measure up to the immense energy challenges the world faces, only one new and clear contender emerges. If we hope to preserve a climate similar to the one found when civilization developed and to which life on Earth is adapted, as well as a stable economy, the time to end energy ignorance is now.

In France, better education has made the public more receptive to nuclear energy. In William Tucker's "Terrestrial Energy," a brochure of a reprocessing center near La Hague reads:

> Source of Life. Since that gigantic nuclear explosion, origin of the universe, called the Big Bang, matter and energy have remained a united and faithful duo. Man himself is stardust... Today the stars, the sun, and the burning core of the Earth are ceaseless beds of nuclear reactions... From distant stars to the Earth's core, it continues its constructive work. Man has learnt to master one nuclear reaction, fission, taming it into a clean and inexpensive energy.

France has today achieved partial energy independence through the use of nuclear fission, and has a third the per capita carbon footprint of the United States. France turned to nuclear energy because of its own paucity of domestic fossil fuel resources, and is now better off as a result. The world must turn to nuclear fission as quickly as possible in the years ahead, following France's example. The source of life, nuclear reactions, must now save the world from the burning of fossilized dead organic matter en masse that is killing the global economy—and the Earth's population. The world must educate its high school science students about energy, and the incredible power of the atom in relation to Einstein's E=MC squared. Education will inspire the young people of the world to enter the metallurgical sciences and become the nuclear engineers that will run the nuclear power plants that will have to provide the major part of our energy needs.

As Admiral Hyman G. Rickover remarked in 1957, "If we give thought to the problem of energy resources, if we act wisely and in time to conserve what we have and prepare well for necessary future changes, we shall insure the dominant position for our own country (19)." Instead, the United States squandered the world's resources and the IFR project that could have saved itself without directing a whim of thought towards the future. It may be too late to ensure that the United States remains a world leader, but if steps are taken to commercially demonstrate IFR technology quickly and deploy it around the world, the United States may once again emerge as a generous nation with leadership on the desperate issues facing the world of the twenty-first century. Energy is the basis of our prosperity. The warnings of Hirsch and Lovelock should not be taken lightly. If action is not taken as quickly as possible, this century will become very different from the last.

WEB REFERENCES

1. Jeff Rubin. "Oil Prices Caused The Current Recession."
   http://www.theoildrum.com/node/4727

2. Faiz Shakir. "Ten Years Ago, Bin Laden Demanded Barrel Of Oil Should Cost $144."
   http://thinkprogress.org/2008/07/05/bin-laden-144-oil/

3. Colin Campbell. "Interview With Colin Campbell."
   http://www.energybulletin.net/node/48713

4. "Shadow Government Statistics."
   http://www.shadowstats.com/

5. Annys Shin. "Stuck at Unemployed: When A Layoff Becomes a Lifestyle."
   http://www.washingtonpost.com/wp-dyn/content/article/2009/06/05/AR2009060503750_2.html?sid=ST2009060501126

6. M. King Hubbert. "Nuclear Energy and the Fossil Fuels."
   http://www.energybulletin.net/node/13630

7. "Peak Oil Primer And Links."
   http://www.energybulletin.net/primer

8. "World Oil Production Forecast—Update May 2009."
   http://www.theoildrum.com/node/5395

9. Michael Klasgen. "The IEA Warns Of Shortages—'The Next Oil Crisis Is Coming.'"
   http://www.energybulletin.net/node/48582

10. Roscoe Bartlett. "Campaign Finance Reform And Peak Oil."
    http://bartlett.house.gov/uploadedfiles/March%2029,%202007.pdf

11. Walter Youngquist. "Alternative Energy Sources."
    http://www.hubbertpeak.com/Youngquist/altenergy.htm

12. James Lovelock. "Nuclear Energy: The Safe Choice For Now."
    http://www.ecolo.org/lovelock/nuclear-safe-choice-05.htm

13. James Hansen. "Tell Barack Obama The Truth: The Whole Truth."
    http://www.columbia.edu/~jeh1/mailings/20081121_Obama.pdf

14. Tom Blees. "Rebuttal To Greenpeace On Nuclear."
    http://www.marklynas.org/2009/1/5/tom-blees-rebuttal-to-greenpeace-on-nuclear

15. Robert Bryce. "The Terawatt Challenge."
    http://www.energytribune.com/articles.cfm?aid=271&idli=1

16. Bernard Cohen. "Breeder Reactors: Renewable Energy Source."
    http://www.sustainablenuclear.org/PADs/pad11983cohen.pdf

17. Alvin Weinberg. "The Second Fifty Years Of Nuclear Fission."
    http://www.cns-snc.ca/history/fifty_years/weinberg.html

18. Steve Kirsch. "The Integral Fast Reactor (IFR) Project."
    http://skirsch.com/politics/globalwarming/ifr.htm

19. Admiral Hyman G. Rickover. "Energy Resources And Our Future."
    http://www.energybulletin.net/node/23151

20. Robert Bryce. "Green Energy Advocate Amory Lovins: Guru Or Fakir?."
    http://www.energytribune.com/articles.cfm?aid=676

21. Marcos Chamon. "China Economics Blog: Cars in China."
    http://china-economics-blog.blogspot.com/2008/07/cars-in-china-more-research.html

22. "Tesla—Model S."
    http://www.teslamotors.com/models/index.php

23. Peter Glover and Michael Economides. "The Alice In Wonderland World Of The Greens."
    http://www.energytribune.com/articles.cfm?aid=1797

24. Ken Zweibel, James Mason and Vasilis Fthenakis. "A Solar Grand Plan."
    http://www.scientificamerican.com/article.cfm?id=a-solar-grand-plan

25. KATU Staff. "Man Killed When Wind Tower Collapses."
    http://www.katu.com/news/9383316.html

26. Larry Bell. "Alternative Energy Opinions: Getting A Real Grip On 'Green.'"
    http://www.energytribune.com/articles.cfm?aid=949&idli=3

27. Christopher Booker. "The E.U.'s Wind Power Self-Deception."
    http://www.energytribune.com/articles.cfm?aid=588

28. "E.ON Netz Wind Report 2005."
    http://www.windaction.org/documents/461

29. Robert Bryce. "The Promise Of Biofuel Is A Lie—Der Spiegel Exposes The Brazilian Ethanol Madness."
    http://www.energytribune.com/articles.cfm?aid=1325

30. Michael Economides. "The Energy Debit Of Making Ethanol."
    http://www.energytribune.com/articles.cfm?aid=93&idli=3

31. Robert Bryce. "The Cellulosic Ethanol Mirage: Verenium And Aventine Are Circling the Drain."
    http://www.energytribune.com/articles.cfm?aid=1506

32. Robert Rapier. "Book Review: Green Algae Strategy."
    http://www.theoildrum.com/node/5465#more

33. Alice Friedmann. "The Hydrogen Economy—Energy And Economic Black Hole."
    http://www.energybulletin.net/node/2401?ENERGYBULL=3f111b51386b890bdf16abbd3c38e150

34. Brad Lemley. "Anything Into Oil."
    http://discovermagazine.com/2003/may/featoil/

35. "If 2% Leaks, The CO_2 Impact Of Natural Gas Is The Same As Burning Coal."
    http://un-naturalgas.org/weblog/?p=198

36. Patrizia Minutolo, et. Al. "Particulate Emission From Natural Gas Burning Home Appliances Is Assessed In Environmental Engineering Science Journal."
    http://www.bio-medicine.org/medicine-news-1/Particulate-Emission-from-Natural-Gas-Burning-Home-Appliances-Is-Assessed-in-Environmental-Engineering-Science-Journal-33399-1/

37. "Oil and Natural Gas Production—Additional Information"
    http://www.epa.gov/air/community/details/oil-gas_addl_info.html

38. D. W. Dixon. "Radon Exposures From The Use Of Natural Gas In Buildings."
    http://rpd.oxfordjournals.org/cgi/content/abstract/97/3/259

39. "Deadly Power Plants? Study Fuels Debate."
    http://www.msnbc.msn.com/id/5174391/

40. S. H. Mohr and G. M. Evans. "Forecasting Coal Production Until 2100."
    http://www.theoildrum.com/node/5256

41. Xina Xie. "Senator Kerry, Here's A Reality Check."
    http://www.energytribune.com/articles.cfm?aid=1871

42. Mara Hvistendahl. "Coal Ash Is More Radioactive Than Nuclear Waste."
    http://www.scientificamerican.com/article.cfm?id=coal-ash-is-more-radioactive-than-nuclear-waste

43. Keith Johnson. "Duke Nuke 'Em: CEO Rogers 'Betting' on Nuclear Power."
    http://blogs.wsj.com/environmentalcapital/2009/05/26/duke-nuke-em-ceo-rogers-betting-on-nuclear-power/

44. "Fast Breeder Reactors."
    http://hyperphysics.phy-astr.gsu.edu/hbase/NucEne/fasbre.html

45. Charles Till. "Plentiful Energy And The IFR Story."
http://www.sustainablenuclear.org/PADs/pad0509till.html

46. George Stanford. "Integral Fast Reactors: Source Of Safe, Abundant, Non-Polluting Power."
http://www.nationalcenter.org/NPA378.html

47. Dirk Kempthorne. "Congressional Record—Charles Till."
http://www.anl.gov/Media_Center/Argonne_News/news97/crtill.html

48. "The U.S. Generation IV Fast Reactor Strategy."
http://www.ne.doe.gov/pdfFiles/genIvFastReactorRptToCongressDec2006.pdf

49. Gerald Marsh and George Stanford. "How to Spread Nuclear Power Without Spreading Nuclear Know-How."
http://www.sustainablenuclear.org/PADs/pad0611marsh.pdf

50. "Senator Craig And Al Gore."
http://www.youtube.com/watch?v=zZdyRl04OVg

51. "Oklo: Natural Nuclear Reactors—Fact Sheet."
http://www.ocrwm.doe.gov/factsheets/doeymp0010.shtml

52. "Carbon Dioxide Created By One Kilowatt-Hour."
http://www.stewartmarion.com/carbon-footprint/html/carbon-footprint-kilowatt-hour.html

53. Bernard Cohen. "Costs Of Nuclear Power Plants—What Went Wrong?."
http://www.phyast.pitt.edu/~blc/book/chapter9.html

54. Barry Brook. "Nuking Green Myths."
http://www.theaustralian.news.com.au/story/0,25197,25607083-7583,00.html

55. Barry Ritholtz. "Fed Financing $18.5 Billion Next Gen Nukes."
http://www.ritholtz.com/blog/2009/06/fed-financing-185-billion-nextgen-nukes/

56. Felix Salmon. "Nuclear Power: Going Fast."
http://blogs.reuters.com/felix-salmon/2009/06/23/nuclear-power-going-fast/#comment-3135

57. Robert Boyar. "Radiation and Common Sense."
http://www.sustainablenuclear.org/PADs/pad9709boyar.html

58. S. M. Javad Mortazavi et al. "Introduction To Radiation Hormesis."
http://www.sustainablenuclear.org/PADs/pad0110mortazavi.pdf

59. S. M. Javad Mortazavi. "Natural Radiation: High Background Radiation Areas (HBRAs) Of Ramsar, Iran."
http://www.angelfire.com/mo/radioadaptive/ramsar.html

60. "FRONTLINE: Nuclear Reaction: Three Mile Island."
http://www.pbs.org/wgbh/pages/frontline/shows/reaction/readings/tmi.html

61. Bernard Cohen. "The Chernobyl Accident—Can It Happen Here?."
http://www.phyast.pitt.edu/~blc/book/chapter7.html

62. Nissa Darbonne. "Valentine's Dinner With Alan Greenspan: Nuclear!."
http://blogs.oilandgasinvestor.com/nissa/2008/02/14/valentines-dinner-with-alan-greenspan-nuclear-plus-iraq-and-are-we-in-a-recession/

63. Stephanie Rogers. "Transition From Oil To Renewable Energy 100 Years Away."
http://www.mnn.com/earth-matters/energy/stories/transition-from-oil-to-renewable-energy-100-years-away-says-exxon-mobil

64. Michael d'Estries. "Survey: Women Fail On Energy Knowledge."
http://www.mnn.com/earth-matters/energy/stories/survey-women-fail-on-energy-knowledge

Glossary

Actinide: An element in the same column on the periodic table as element #89, actinium. Includes thorium, uranium, plutonium, and higher transuranics like americium, which are bred during fission reactions from fertile isotopes. In the case of IFR fuel, the presence of Pu-240 mixed in with Pu-239 makes centrifugal purification of Pu-239 to weapons-grade levels extremely difficult due to the similar mass of the two isotopes.

Alpha radiation: Positively charged particle emitted by elements like uranium 238 or plutonium 239. It is composed of two protons and two neutrons.

Anthracite: A high-grade form of coal with an energy density as high as 30 million BTUs per ton.

Anthropogenic: Of human origins; often used to describe global warming as being caused by human activity rather than natural factors.

Argonne National Laboratory: Located in Illinois, ANL is one of nine national laboratories under the aegis of the DOE. Developed the Integral Fast Reactor from 1984 to 1994.

Atomic energy: Energy on the order of two million times the density of chemical energy, released when a neutron collides with the nucleus of an atom, causing it to fission and 0.09 percent of its mass to be converted to pure energy according to the formula $E = MC^2$.

ALMR: Advanced Liquid Metal Reactor. Another name for the Integral Fast Reactor, which uses liquid sodium metal as its heat transfer medium, as opposed to pressurized water.

Background radiation: Natural radiation all around us, mostly from radon gas emanating from the Earth, cosmic rays and food. The average American is exposed to about 360 millirem per year.

Base-load: Power reliably supplied to the grid 24/7 as demand dictates. Base-load power plants are powered by nuclear, fossil fuels, or hydro, since these energy sources use stored energy that can be released as desired. Wind and solar cannot produce base-load, reliable power.

Becquerel: Radioactivity equal to one disintegration per second.

Beta radiation: A negatively charged particle consisting of one electron, emitted by radioactive decay. Unlike alpha particles, beta particles can penetrate human skin.

Bitumen: A black, oily, viscous hydrocarbon. Mixed with minerals, it forms asphalt.

Bituminous: A higher grade of coal with a similar energy density to anthracite, but often with more impurities.

BN-600: A fast breeder reactor that has reliably provided commercial power to the grid in Beloyarsk, Russia since 1980.

Boron: The fifth element on the periodic table. If burnt, boron can be used as an energy carrier. This involves driving oxygen off the rust byproduct for recycle.

Breeder reactor: A nuclear reactor configured so as to produce more fissile material than it consumes. Also called "fast reactor" because of its fast neutron spectrum.

BTU: British Thermal Unit. A measure of energy content.

Capacity: A maximum amount of electric power that can be generated.

Capacity factor: The amount of overall power produced relative to the theoretical maximum. Solar has a capacity factor of 14 percent because the sun isn't powerful most of the time, so one gigawatt of solar capacity would actually only produce a sporadic 140 megawatts of power. A one-gigawatt nuclear plant, in contrast, produces close to one gigawatt of reliable power since it can run 90 percent of the time, stopping only for routine maintenance. As of 2008, there were only 10 square miles of solar panels in the entire world, with 5,000 megawatts of capacity. However, since their capacity factor is low, all the solar panels in the entire world actually produce less than a single GE-Hitachi boiling water reactor rated at 1,100 megawatts.

Carbon tax: Any taxation leveled at hydrocarbon based fuels, with the goal of reducing greenhouse gas emissions by making fossil fuels more expensive, and helping to raise funds to subsidize alternatives. In Europe, carbon taxes have thus far failed to lower emissions of greenhouse gases. Only nations that rely heavily on nuclear and hydropower have the lowest emissions, such as France and Sweden.

Carrier: A material that carries energy from point a to point b, such as electricity, hydrogen, boron, or gasoline. Uranium can only directly power very large vessels such as submarines, but electricity from nuclear central stations can carry energy to home heaters, appliances and electric cars. Carriers cannot be used to obtain energy based upon the second law of thermodynamics, although some sources of energy are also carriers as well, like oil.

Carrying capacity: The measure of the size of a population an ecosystem can support. Humans have increased the carrying capacity of their environment by domesticating plants and animals for food, and by using fossil energy to further industrialize agriculture.

Centrifuge enrichment: A method of uranium enrichment wherein isotopes of gaseous uranium 238 and 235 are separated from one another based upon their weight, when the centrifuge spins at high speed to cause the heavier isotopes to move farther. The process is up to fifty times more energy efficient than diffusion enrichment.

Cs-137: A beta-gamma emitting fission product with a half-life of 26.6 years.

Closed fuel cycle: A fuel cycle where created actinides are sent back into the reactor to be split apart, thereby releasing more energy. The process is over 100 times as fuel efficient as a once-through fuel cycle.

CRBR: Clinch River Breeder Reactor. A large-scale commercial breeder planned for the Clinch River, Tennessee. It was cancelled in 1983 due to proliferation concerns.

CO_2: Carbon dioxide, the greenhouse gas most responsible for climate change, emitted by the burning of fossil fuels.

Demand destruction: a fall in demand for a commodity, such as oil, resulting from very high prices. When oil prices reached $147 a barrel, this lowered demand for oil and the services it provides such as trade and transport, killing the economy. The resulting drop in demand also lowered oil prices. As the economy recovers, oil prices will rise.

Depleted uranium: Uranium 238 separated from uranium 235 by an enrichment process. This can be a fuel for fast reactors, which use it to produce plutonium.

Deuterium: An isotope of hydrogen composed of one proton and one neutron. It is a potential nuclear fusion fuel.

Diffusion enrichment: An older, energy-intensive method of uranium enrichment where super-heated gaseous uranium is sent through a porous membrane to separate the heavier U-238 out from the lighter U-235.

DOE: Department of Energy; its stated goal is the advancement of national, economic, and energy security for the United States.

EBR-II: Experimental Breeder Reactor II. A breeder reactor prototype used to demonstrate the Integral Fast Reactor concepts of complete reprocessing with metal-alloy fuel, passive safety, and proliferation-resistance. The EBR-I was an earlier version from the `50s.

Electron: A tiny negatively-charged particle with 1/1837 the mass of a proton. A stream of these particles can transfer energy in the form of electricity.

Electrorefining: A pyrometallurgical refining process that involves oxidation of metallic impurities in a high temperature liquid bath, then uses the positive and negative charges of an electrolytic cell to attract dissolved metals to different anodes. The deposition of metals is then in a purer form. The process is used in an IFR to self-reprocess transuranic actinides for recycle. The process is well understood in industry, but still needs to be demonstrated in a large-scale commercial Integral Fast Reactor in order for an IFR to receive approval for construction by the NRC.

Emission: A discharge of pollution directly into the environment.

EROEI: Energy Return On Energy Invested. It always takes some energy to obtain energy, usually in harvesting it and refining it. If the EROEI is 1:1 or less, the process ceases to be a source of energy. For instance, oil gushing out of the ground may have an energy return of 100:1, but by the time most has been extracted the EROEI falls to 1:1 as sucking the rest out of the ground becomes too difficult, and production of oil ceases.

Fallout: Often radioactive particles that result from a nuclear explosion and descend through the air. This has nothing to do with nuclear power, and should be associated with bombs.

Fast breeder reactor: A reactor configured so as to produce more fissile material than it consumes. Using low-grade ores, this type of reactor has an unlimited fuel supply.

Fast neutrons: Neutrons with kinetic energy on the order of millions of electron volts. These neutrons are slowed in thermal reactors like the LWR, but are kept fast to breed more fuel in fast reactors.

Fast reactor: A nuclear reactor that uses fast neutrons given off during fission to breed fissile material from fertile material. Since 99.3 percent of uranium is fertile and 0.7 percent is fissile, these reactors are over 100 times as fuel efficient—the key to exploiting the exponentially more abundant lower grade ores economically and having an all but inexhaustible fuel supply. Generation IV nuclear reactor designs include fast reactors.

Fertile: An element that undergoes radioactive decay to become fissile when it absorbs a neutron, such as U-238 or Th-232, which become Pu-239 or U-233, respectively.

Fissile: An element capable of sustaining a fission chain reaction when split, such as Pu-239 or U-235. U-235 is the only fissile material found in nature. The more widely abundant U-238 and Th-232 are fertile, and can be used to breed fissile material.

Fission: A nuclear process in which heavy, unstable elements are split to form lighter elements, releasing great quantities of heat.

Fission products: A new atom created by fission of a heavy atom. This doesn't include new elements bred from fertile isotopes like Pu-239, which are actinides.

First law: Of thermodynamics. Known as the law of conservation, the amount of energy in the universe is fixed. When energy is released from matter as a physical phenomenon, it simply converts from a concentrated state to a less concentrated state. Nothing new can be created that wasn't around in the first place.

Fuel cycle: The entire set of stages involved in the utilization of nuclear fuel, including mining, milling, enrichment, transportation, reprocessing, and waste disposal. The Integral Fast Reactor is designed to reduce the number of these steps, eliminating some entirely, such as mining, milling and enrichment.

Fusion: A phenomenon that occurs when great heat and pressure overcome the positive repulsion of the nuclei of atoms, allowing them to get close enough together for the strong force to fuse them together. A small amount of mass is converted into energy in the process.

Gamma radiation: Very penetrating, high-energy electromagnetic radiation emitted from the nucleus of an atom that can only be stopped by materials like lead or thick concrete.

Gas-Cooled Fast Reactor: An innovative generation-IV nuclear fast reactor design cooled by helium gas rather than liquid metal. Thermal efficiencies in this type of reactor may approach 70 percent, compared to around 40 percent in liquid metal reactors. This type of reactor will likely be commercialized after 2030.

GDP: Gross Domestic Product. This measure of economic activity is heavily dependant upon energy consumption.

Gen-III+: Generation three plus light water reactors. Innovative new reactor designs improving upon 1990s Generation-III plants, with improvements in safety, economics, and efficiency. These plants are now being built around the world, with the first scheduled to open in 2012 in Olkiluoto, Finland. Over 100 are proposed for use in the United States, and over thirty are in the early stages of construction or planning.

Gen-IV: Generation IV nuclear energy plants, designed to completely solve the major problems associated with nuclear: waste, cost, safety, proliferation, and fuel supply. These reactors are often fast-spectrum, meaning they are improved breeder reactors, but can also operate in burner mode. The Integral Fast Reactor was among the first of this type of reactor.

Gigawatt: (GW). One billion watts.

Greenhouse gases: Gases that trap the heat of the sun by preventing a portion of the sun's rays from reflecting back out into outer space. CO_2 is the major greenhouse gas.

Grid: The layout of an electrical distribution system.

Half-life: The time it takes for half the original amount of a radioactive substance to disintegrate. Half of the remaining portion will disintegrate in the same fixed period of time, and so on.

Holocene: the period of mild climate that has blessed the Earth for the past 10,000 years, and has been with us since the beginnings of civilization. It likely triggered the rise in agriculture that allowed humans to begin to climb White's ladder, and may no longer be with us unless coal plants are linearly phased out over the next two decades.

Homeostasis: An organism's maintenance of a relatively stable state of internal order, in terms of body heat and chemical concentrations. This requires energy.

Hormesis: The biological theory that organisms are made more resilient by exposure to substances in low doses that may still be harmful in high doses.

IGCC: Integrated Gasification Combined Cycle. A state-of-the-art coal plant that burns coal in the form of a syngas at a higher thermal efficiency than pulverized coal plants, while also capturing particulates and carbon dioxide. Large-scale, long-term storage of the waste is the problem, since millions of tons are produced every year.

Integral Fast Reactor: A revolutionary Generation IV nuclear plant designed at Argonne National Laboratory in the decade from 1984 to 1994. The reactor was cooled by liquid sodium metal, and used an on-site, integrated fuel cycle that segregates and consumes plutonium at the same rate it is created for over 100 times the energy per ton of uranium, and mixes materials in such a way that they are dangerous to handle and not suitable for bomb production. The reactor was designed to be inherently safe, and burn existing stockpiles of nuclear waste as its fuel source. It was cancelled in 1994 for political reasons. In 2001, as part of the Generation IV roadmap, the DOE tasked a 242 person team of scientists from the DOE, UC Berkeley, MIT, Stanford, ANC, LLNL, Toshiba, Westinghouse, Duke, and the Electric Power Research Institute to evaluate 19 of the best reactor designs on 27 different criteria. The IFR ranked #1 in their study which was released April 9, 2002.

International Energy Agency: (IEA). A Paris-based energy research organization. The IEA has become increasingly alarmed by the energy problem after its 2008 World Energy Outlook report which determined that 580 of the 800 largest oil fields in the world are in decline, and they may be declining at faster rates than once expected. In its report, the IEA said "it is not an exaggeration to claim that the future of human prosperity depends on how successfully we tackle the two central energy challenges facing us today: securing the supply of reliable and affordable energy; and effecting a rapid transformation to a low-carbon, efficient and environmentally benign system of energy supply. What is needed is nothing short of an energy revolution."

I-129: A very weakly radioactive, long-lived fission product that emits beta particles and gamma rays. It has a half-life of 15.7 million years.

Isotope: Atoms of the same element that have an equal number of protons, but a different number of neutrons. For example, uranium 238 and uranium 235, or plutonium 239 and plutonium 240.

ITER: International Thermonuclear Experimental Reactor. A nuclear fusion power plant scheduled to open in 2018, which will produce net energy from fusion in the form of heat, a first for fusion power plants.

Jevons Paradox, the: A theory of economics that efficiency increases energy consumption. It will likely remain true until a majority of the world's population owns a car. Efficiency just can't keep up with exponential demand growth.

Kerogen: A mixture of organic chemical compounds that make up a portion of the organic matter found mixed in with sedimentary rock. Mostly fossilized algae and woody plant remains, this forms the hydrocarbon basis of shale oil. Extracting it and refining it is extremely damaging to the environment and energy intensive, and most likely will require more energy than it yields.

Kilowatt: A thousand watts.

Kilowatt-hour: A measure of electricity consumption used as a unit of work or energy, defined as 1,000 watts of power expended over one hour. One kilowatt-hour is equal to 3,412 BTUs. The average American household uses about 38 kilowatt-hours a day, and this will increase as more electric space heaters and electric cars are deployed.

Light water reactor: (LWR). The most common type of nuclear plant used around the world, cooled by pressurized water, which also acts as a neutron moderator. LWRs use less than 1 percent of the potential energy of the uranium, whereas fast reactors use 99.5 percent of it. There is a heavy water reactor, used in Canada under the name CANDU. This type of reactor is cooled by water that contains the hydrogen isotope deuterium.

Lignite: A grade of coal with about one quarter to half the energy density of bituminous coal that often appears brown in color.

Liquid Fluoride Thorium Reactor: (LFTR). An innovative thermal-neutron generation IV nuclear energy plant that uses U-233 as fuel absorbed in a fluid salt solution. The reactor breeds as much U-233 from Th-232 as it uses. Like the IFR, the reactor will mix materials in such a way that they are dangerous to handle and not suitable for bomb production. However, the LFTR still requires more R&D than the IFR before it will be ready for commercialization. It also has a low breeding ratio, around 1.01. This means it will take a long time for such a reactor to breed enough extra fuel to start up another LFTR.

LNG: Liquefied Natural Gas. Gas is much more difficult to transport overseas than oil, and requires this process to become dense enough to be imported overseas.

Load-cycling: The rapid fluctuation of a base-load power plant's power output to help balance out sporadic and unreliable output from intermittent power sources, like wind. Only hydro dams or special gas or oil-fired combustion engines are able to deal with these fluctuations. Such combustion engines are also less efficient, supplementing over consumption of fossil fuels and offsetting wind's contribution, and they are less resilient and wear out faster.

Mbpd: Million barrels (of oil) per day.

Megawatt: (MW). A unit of electrical power equal to 1,000 kilowatts.

Metallurgical: Of or relating to the science and technology of metals.

Methane: Another name for natural gas. Methane is colorless, odorless and flammable, although gas companies mix in materials with the gas that give it an odor for safety purposes. Methane is twenty-three times as potent a greenhouse gas as CO_2.

Millirem: A unit of absorbed radiation dose equal to 0.001 rem. Eating a banana will expose someone to 0.01 millirem, whereas living near a nuclear plant will expose someone to 0.009 millirem annually.

Moderator: Any material that can slow down neutrons. Water is the moderator for most thermal reactors, and acts by slowing down neutrons to allow fission of primarily U-235.

MOX: Mixed oxide fuel. A reactor fuel created from used nuclear fuel by mixing Pu-239 and U-235 recovered from spent fuel with U-238.

Neutron: A particle with no electrical charge that helps make up the nucleus of an atom. Under proper conditions, the neutral charge of a neutron allows it to be absorbed by fissile nuclei, causing them to become unstable and then undergo fission.

Neutron poison: A material with a large neutron absorption cross-section, when slowing neutrons is an undesired effect.

NRC: The Nuclear Regulatory Commission. Insures safety of nuclear plants and the use of radioactive materials in medical, industrial, and research applications.

Nuclear Battery: Name for a small, self-contained, sealed nuclear power plant that can supply power to people living off the grid. Toshiba is now marketing its "4S," Super-Safe, Small and Simple.

NYMEX: New York Mercantile Exchange. Prices for oil, natural gas and other commodities are determined through this exchange by traders based upon supply and demand factors.

Peak-load: Power put into the grid when demand peaks in the middle of the day, which solar can help provide since it works efficiently near noon.

Peak oil: The point when oil extraction reaches a maximum then goes into inexorable decline. This can refer to a single oil well, a nation, or the entire world's oil production. The world almost certainly peaked in July 2008 at 87.9 mbpd.

Photovoltaic panel: A panel often made from silicon that generates electricity upon contact with light.

Plasma: Superheated material that has been stripped of its electron cloud, and under high enough temperatures and pressures can fuse to form new elements and release energy. This is the fourth state of matter after gas, liquid, and solid.

Positive feedback loop: Any self-reinforcing process that can run away like a domino effect. For example, if global warming gets much worse, frozen methane could be released, which will greatly increase warming, melting even more methane. Or, if the graphite of an RBMK reactor heats up, fission occurs faster, heating up the reactor to cause fission to go still faster.

Positive temperature coefficient: A characteristic of RBMK nuclear plants such as the one in Chernobyl, where if the reactor overheats fission occurs more quickly.

Positron: The antimatter equivalent of an electron.

Ppm: Parts per million.

Pressurized water reactor: Any reactor cooled by pressurized water. This includes light water reactors and CANDU heavy water reactors, which use water containing deuterium and require uranium of low or no enrichment.

PUREX: Plutonium and Uranium Recovery by EXtraction. This process is used to isolate purified plutonium from LWR spent fuel to produce MOX fuel for thermal-neutron reactors. While the plutonium is of sufficient weapons-grade, it is of poor isotopic quality due to the trace presence of Pu-240 and Pu-241.

Pyrometallurgical: An ore-refining process of electrorefining.

Pyroprocessing: The name for the pyrometallurgical process of electrorefining used in the IFR fuel cycle.

Radiation: Any high-speed transmission of energy in the form of particles of electromagnetic waves.

Ratcheting: A process of greatly increasing the materials required to construct a nuclear plant, which may reach the point where it actually decreases safety of a plant by making it too complicated.

RBMK: Reaktor Bolshoy Moshchnosti Kanalniy, or "High Power Channel Type Reactor." A Russian graphite-moderated thermal-neutron reactor with a high positive temperature coefficient, also known as void coefficient, designed for simultaneous power generation and plutonium production for bombs. If the reactor overheats, positive feedback loops kick in that encourage fission to speed up faster and faster. 12 RBMK reactors are still in use in Russia and Lithuania. They are probably safer than coal.

Reactor years: Operational years of performance of nuclear reactors. For instance, a thousand reactors running for ten years would be ten thousand reactor years.

Rem: Roentgen equivalent man. A unit measure of absorbed dose of radiation, taking into account the quality factor of the particular radiation.

Reprocessing: Chemical treatment of spent nuclear fuel that separates unused uranium and plutonium from fission products. This allows much more energy to be obtained per ton of uranium, while also reducing the volume and toxicity of the waste.

Second law: Of thermodynamics. Known as the law of entropy, energy will always tend to move from a concentrated state to a less concentrated state. As it moves from one state to the other, it can do work. No energy, once used, can be regenerated or be "renewable."

Sodium: The eleventh element on the periodic table, which has excellent heat-transfer characteristics and is used as the coolant for the IFR. Sodium is a metal at room temperature, but becomes a liquid at 207°F.

Sodium-Cooled Fast Reactor: The new name for the Integral Fast Reactor under the current Generation IV program, with some tiny improvements thrown in. This type of fast reactor increases the efficiency of uranium resource utilization by a factor of 160.

Spent fuel: Irradiated reactor fuel that is no longer capable of sustaining a fission chain reaction, due to the buildup of neutron poisons. In the United States and in Sweden, spent fuel is stored away and called "nuclear waste." Elsewhere, it is reprocessed for more energy using PUREX.

S-PRISM: The safety design used by the IFR that employs passive safety characteristics. Even if thousands of IFRs were built, the odds of a meltdown would be around once every 400,000 years.

Sr-90: A highly radioactive fission product with a half-life of just twenty-eight years. It is a hard beta emitter.

Sub-bituminous: A grade of coal with about one third to two-thirds the energy density of bituminous coal.

Super Phoenix: (French Super Phenix). The largest fully commercial fast breeder reactor ever built, near the Rhone River in France. The Super Phoenix provided reliable power to the grid at less than twice the cost of LWRs since 1985, but was shut down for political reasons in 1997.

Tc-99: A weakly radioactive fission product found in IFR spent fuel that emits beta particles. It has a half-life of 211,000 years.

Terawatt: (TW). One trillion watts.

Thermal efficiency: The measure of how much useful work is extracted from a given source of energy. For internal combustion engines, it is about 30 percent, meaning 30 percent of the energy of the fuel does work and 70 percent is lost as waste heat due to inefficiencies of the second law.

Thermal neutron reactor: Another name for a LWR or any reactor using slowed neutrons, such as the LFTR or even the Soviet RBMK design of

Chernobyl. These reactors slow their neutrons using a neutron moderator. They are the opposite of a "fast reactor," but are not commonly called "slow reactors."

Thorium: A fertile heavy metal four times as abundant as U-238 in the Earth's crust, which can be used to breed uranium 233. India has large reserves, and is pursuing this type of reactor. It is less abundant than uranium in seawater.

Transuranic: An artificially made, radioactive element that has an atomic number higher than uranium on the periodic table, such as plutonium or americium.

Tritium: A radioactive isotope of hydrogen composed of one proton and two neutrons. It has a half-life of 12.4 years, and is a potential nuclear fusion fuel.

U-3: Measure of official unemployment rate used by the U.S. Bureau of Labor Statistics. As of July 2009, this unemployment rate was 9.5 percent.

U-6: U.S. Bureau of Labor Statistics unemployment measure that counts "discouraged workers" who have stopped looking for work because they have had difficulty finding work due to economic factors, "marginally attached workers" who have not looked for work recently but would still like to work and occasionally seek work, and part time workers who cannot find full employment. As of July 2009, this unemployment rate was 16.5 percent.

Unconventional oil reserves: Nonliquid oil reserves such as the Canadian tar sands and shale oil in the American west. Producing these reserves is extremely energy intensive, and extraction rates tend to be much slower than those of conventional reserves.

Uranium: A naturally occurring, silvery heavy metal that fuels nuclear power reactors. U-238 is the dominant isotope, which is fertile and can be used to produce plutonium. 0.7 percent of uranium is U-235, the fissile isotope used to fuel LWRs. Uranium is as abundant in the Earth's crust as tungsten or tin, and exists in virtually limitless quantities in seawater.

Vitrification: The process of incorporating hazardous material in glass, where it remains immobile and will be shielded from interaction with the environment.

Yellowcake: The nickname for mined uranium. This compound, U3O8, appears yellow in color.

Yucca Mountain repository: An expensive nuclear spent fuel repository cancelled by the Obama administration in 2009. It was decided by secretary of energy Steven Chu that using used reactor fuel in fast-neutron reactors would be a wiser choice. I agree completely.

Index

A

actinide, 155–56
AIG (global insurance corporation), 17
algae, 78–79
Angels & Demons, 92
antimatter, 91–92
  positron, 91, 163
Appel, Brian, 89
atomic energy, 50, 52
Aventine, 78

B

Ballard, Geoffrey, 80
  hydrogen fuel cell, 80
Bartlett, Albert, 35
Bartlett, Roscoe, 26
base-load, 31, 59, 62
Beverly Hill Billies, 20
Bin Laden, Osama, 14
biomass, 19, 40, 75
bitumen, 26
boron, 82
Bottomless Well, The (Huber and Mills), 55

breeder reactor, fast breeder reactor, CRBR (Clinch River breeder reactor), 100, 155
Brown, Lester R.
  "Why Ethanol Production Will Drive World Food Prices Even Higher in 2008," 77
BTU (British thermal unit), 76

C

Campbell, Colin, 11, 16
capacity, 59, 65–66, 69
  kilowatt, 59, 66, 72
  kilowatt-hour, 59, 61, 72
capacity factor, 59
carbon tax, 100
carrier, 14, 56, 82
carrying capacity, 50, 76
China, 31, 47, 56, 73, 77, 85, 97–98
  Three Gorges Dam, 73
  Yangtze River, 73
civilization, 9, 33, 42, 47, 58, 65
Clausius, Rudolf, 38
  entropy, 38, 40, 80
climate change, anthropogenic, 29–31, 97, 99

Clinton, Bill, administration, 17, 81
CO_2, 26, 30–31, 78, 93–94, 98–99
coal, 13, 25, 31–32, 37, 45, 56, 61, 66, 68, 93, 96–97
  anthracite, 96
  bituminous, 96
  lignite, 68, 96
Cohen, Bernard, 44

D

Darwin, Charles, 50
  theory of evolution, 50
Darwin, Charles Galton
  Next Million Years, The, 50
Denmark, 67
deuterium, 86–87
Diamond, Jared
  Guns, Germs, and Steel, 42
DOE (Department of Energy), 91
domesticated animals, 41
domesticated crops, 43, 50
Dow Jones, 17
Drake, Edwin, 19

E

"Eating Fossil Fuels" (Pfeiffer), 76
EBR-I (experimental breeder reactor), 44
Economides, Michael J., 76
economy, 9, 13–18, 22, 48, 75, 84
Edison, Thomas, 55
efficiency of energy, 23, 32, 44, 49, 51, 83–85
Einstein, Albert, 49
Electric Car, 59
electricity, 14, 30, 55–57, 62–63
electron, 60
EROEI (energy return on energy invested), 14, 22

ethanol, 75–78
  cellulosic, 78
  corn, 75–77, 79
  sugarcane, 75
evolution, 41, 51. *See also under* Darwin, Charles

F

fast neutrons, 87
  I-129, 160
fast reactor, 100
  integral, 23, 32
  sodium-cooled, 44
Fertile Crescent, 42–43
fissile material, 45
fission, 42, 45, 56
fossil fuels, 13, 23, 39, 41–42, 47, 49–50, 59–60, 86
Friedman, Thomas
  Hot, Flat, and Crowded, 13, 56
fuel cycle
  closed, 53
  pyrometallurgical, 32
fusion, 40, 86–87, 91

G

gasoline, 14, 19–20, 60
  E10, 78
  E25, 75
GDP (gross domestic product), 16, 84
Gen-III+ (Generation three plus light water reactors), 32
Gen-IV (Generation IV nuclear energy plants), 158
GM (General Motors), 17
Gram, Niels, 67
Grand Inga Dam, 73
greenhouse gases, 29–31
  carbon dioxide, 30

nitrous oxide, 77
grid, 59, 61–62, 69
Guns, Germs, and Steel (Diamond), 42
GW (gigawatt), 73, 98

H

half-life, 87
Hansen, James
   "Tell Barack Obama the Truth: The Whole Truth," 30
Hirsch, Robert L., 27
   Hirsch report, 27
Holocene, 30, 42, 159
homeostasis, 38, 40
Hot, Flat, and Crowded (Friedman), 56
Hubbert, M. King, 20–21, 23, 27, 51
Huber, Peter
   Bottomless Well, The, 55
hydrogen, 9, 27, 40, 46, 76, 80–82

I

IEA (International Energy Agency), 25
IFR (integral fast reactor). *See under* fast reactor
IGCC (integrated gasification combined cycle), 97–98, 100
India, 47, 56, 88, 97
Industrial Revolution, 43
IPCC (Intergovernmental Panel on Climate Change), 30
isotope, 44, 86
ITER (International Thermonuclear Experimental Reactor), 87

J

Jantsh, Erich, 41
   Self-Organizing Universe, The, 41
Jevons, William Stanley, 83

Jevons paradox, 83–85

K

kerogen, 27

L

Lehman Brothers, 17
LFTR (liquid fluoride thorium reactor), 32
load cycling, 66–68
Lovelock, James, 29, 99
Lovins, Amory, 84

M

Malthus, Thomas, 50
mbpd (million barrels per day), 21, 25, 84, 162–63
methane, 31, 93–94. *See also* natural gas
methane hydrates, 88, 97
Mills, Mark
   Bottomless Well, The, 55
MW (megawatt), 46, 69–70, 87

N

NASA (National Aeronautics and Space Administration), 30, 91
natural gas, 13, 25–26, 80–81, 88, 93–94
   LNG (liquefied natural gas), 21, 95
neutron, 86
neutron poison, 162, 165
Nevada Solar One, 63
Next Million Years, The (Darwin), 50
Norway, 24–25, 67, 72
nuclear power, 10, 26–27, 30, 32, 44, 51
nuclear waste, 30, 100
NYMEX (New York Mercantile Exchange), 15

O

oil, 13–15, 19, 21, 26, 76, 95
  cost of, 15, 17
  diesel, 75–76
  peak, 19, 21–22
oil field, 14, 19–20, 22, 25
  Canada, 95
  Cantarell, 24
  Ghawar, 24–25
  Middle East, 21, 24
  North Sea, 24

P

Patzek, Tad W., 76
peak-load, 59
Pfeiffer, Dale Allen
  "Eating Fossil Fuels," 76
photovoltaic panel, 59
plasma, 87
positive feedback loop, 31, 99
power
  geothermal, 39, 70–71, 84
  hydroelectric, 47, 72–74, 83, 100
  solar, 30–31, 39, 47, 58–63, 78, 86, 91
  tidal, 69
  wave, 69, 71
ppm (parts per million), 30–31
Pressurized water reactor, 163
proliferation, nuclear weapons, 10

R

radiation, 99
  background, 99
  hormesis, 159
reactor years, 46
Repower America, 62, 65
Rickover, Hyman G., 46

Rocky Mountain Institute, 84
Rogers, Jim, 100
Roosevelt, Franklin, 73
Roosevelt, Theodore, 73
Rubin, Jeff, 15

S

safety, 10
Schopenhauer, Arthur S., 9
Self-Organizing Universe, The (Jantsh), 41
shale oil, 26–27
Simmons, Matthew
  Twilight in the Desert, 22, 25
Simon, Julian, 47
Solar Electrical Generating System, 62
space-based solar arrays, 91
spent fuel, 32, 98
steam engine, 38, 43
Super Phoenix (French Super Phenix), 23, 165
Sweden, 68

T

tar sands, 26
"Tell Barack Obama the Truth: The Whole Truth," (Hansen), 30
Tesla Motors, 57
thermal depolymerization, 89–91
thermal efficiency, 43, 70, 85
thermodynamics
  first law of, 37–38
  second law of, 38, 70
thorium, 23, 32, 45, 47, 86, 99
Three Mile Island accident, 66
tritium, 86–87
TW (terawatt), 28, 32, 49, 58
Twilight in the Desert (Simmons), 22

U

U-3 (official unemployment rate), 17–18
U-6 (U.S. Bureau of Labor Statistics unemployment measure), 17
UK (United Kingdom), 24
unconventional oil resources, 24
United States, 9, 13–14, 17, 19–21, 27, 31, 49, 51, 60, 72–75, 84, 96–97, 100
uranium, 14, 23, 44–45, 47, 59, 99
USS *Nautilus*, 45

V

Verenium, 78
Verne, Jules, 45

W

Weinberg, Alvin
  Second Fifty Years of Nuclear Fission, The, 44
wheat, 42
White, Leslie, 41, 45, 47
White's Law, 41, 84
"Why Ethanol Production Will Drive World Food Prices Even Higher in 2008" (Brown), 77
wind, 47, 58, 61, 65–68
"Wind Report 2005," 67
WWII, 45

Y

yellowcake. *See* uranium

LaVergne, TN USA
26 October 2009

161982LV00005B/63/P